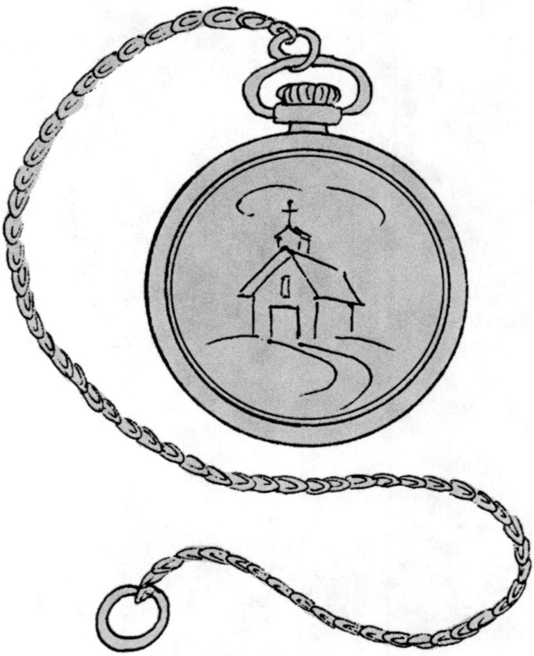

For Maryann,
With thanks for your
enthusiastic support,
Virginia

Amorette in Chicago, about 1890

AMORETTE'S WATCH

A CIVIL WAR WIDOW AND HER GRANDDAUGHTER

by

Virginia Foote

REGENT PRESS
Berkeley, California

ISBN 13: 978-1-58790-164-5
ISBN 10: 1-58790-164-1
LCCN: 2008944275

Manufactured in the U.S.A.
REGENT PRESS
2747 Regent Street
Berkeley, California
www.regentpress.net

Dedicated to my father,

Dellizon A. Foote,

who kept his mother's memory alive

Table of Contents

Part Four: Widow and Son

Part Five: "All the Word 'Mother' Implies"

List of Illustrations

Acknowledgments

To the many people—those I've never met, those I knew in their professional capacity, those I've known for many years—I want to express my deep gratitude for helping me bring this book into being.

I thank Stuart Butler, senior archivist at the National Archives, for a personal letter enclosing the carded medical record that documented Seth Foot's battle wound. James Ogden, historian for the Chickamauga-Chattanooga National Military Park, sent me copied pages from a medical history of the Civil War that I did not know existed. Richard E. Jensen, senior research anthropologist for the Nebraska State Historical Society, contributed facts and wisdom from his wide knowledge of the lands the goldseekers traveled. Becky Wadian, director of library services for Upper Iowa University, sent me my father's college records and introduced me to the story of Susan Collins.

Stephen Yale, archivist for the Northern California Conference of the Methodist Church, directed me to the Claremont School of Theology, where assistant archivist Kim

Creighton showed me the issues of the Southern California Conference Journal that contained all of Mrs. A. E. Foote's annual reports. (I believe it was Amorette who added the *e* to the family's name.) Tracey Del Duca, associate archivist at the General Commission on Archives and History of the Methodist Church (GCAH), supervised the copying and mailing of the photograph of the Los Angeles Conference Deaconess Home that appears in these pages. Jocelyn Rubinetti, Methodist library associate at the Drew Library (where the GCAH is located), found and mailed to me a very useful description of the Chicago Training School.

The library services of the University of California, Berkeley, and in particular the Bancroft Library, with its specialization in Western history, brought to light more than one episode in Seth's story. The Graduate Theological Union Library, also in Berkeley, was my best source of information on the Chicago Training School. I thank this splendid library for furnishing the photograph of the resident faculty. Lynn La Fleur, the highly efficient and unfailingly cheerful director of interlibrary loan services at California State University, Hayward (where I used to teach) gave my requests for obscure periodicals of the nineteenth century personal attention. Professor Mary Cullinan and Professor Eileen Barrett, both of whom were at Hayward (now "East Bay") when I was there, lent their support to my project.

Several Iowa women responded generously to my requests for aid. Ruth Brooks of the Fayette County Historical Society went out of her way to discover the site of Amorette's house in Fayette. Shirley Smith helped me locate an old diary, which in turn helped me locate my grandfather. Linda Sedlmayr, city clerk of Farmersburg, devoted most of her day to driving me first to the farm the Foot family owned and then to the cemetery where Seth was buried.

As the book began to take shape, my four children were among its earliest readers and supporters. John liked the "inventions," the portions in which I imagined a scene (and gave notice to the reader that it was invented). Katherine liked the stories I told about my own life. Rob, speaking as one artist to another, cheered me on. Anne, from her vantage point as a writer and teacher, did me the immense favor of reading the entire manuscript and making intelligent criticisms that guided my final revision. On the technical side of things, I thank my stepson David Anderson, a computer scientist, for twelve years of steady support, ranging from purchasing and installing equipment to patiently coaching me in basic procedures and occasionally rescuing me from untoward consequences.

Two men who were my teachers at writers' conferences made remarks that I have kept in my mind ever since. Douglas Bauer, writer of both fiction and non-fiction, helped me discard theory, look inside myself and find my way. Roy Parvin, a self-described history major who later turned to fiction writing, told me he liked my Missionary Ridge chapter very much, except for the ending. I reread the original ending, saw that it sounded frivolous and removed the frivolity. Two women with specialized knowledge of portions of my story generously read those portions. Mrs. Juliette Smith, librarian at Talladega College after the death of Dr. Helen Bush Caver, made corrections from her firsthand knowledge of college history and local geography. Mary Agnes T. Dougherty suggested some changes in the chapter on the Chicago Training School. Early in my investigation Dr. Dougherty had searched her records and sent me copies of all the clippings she had on Mrs. A. E. Foote.

To fellow members of the Writers' Workshop of the University [of California] Section Club, as well as its offshoot the non-fiction group, I extend my gratitude. I give special thanks to Sally Tubach for her informed criticism, Diane DePisa for her searching questions and Doris Fine for a timely and very welcome piece of advice. I thank Sue Austin, past president of the Workshop, for bringing her organizational skills to bear on questions of publication. I also thank Mary Parks and Dorothy Snodgrass and Phyllis Smith and Deborah Whitney and Mary Kent and Trudi Frei and Joanne Lagerstrom and Frances Townes for advice given and opinions shared.

To my husband, Henry Anderson, I am grateful for so many things. He provided the impetus for my research in the first place; he drove me up the steep slopes of Missionary Ridge; he carefully edited the chapters I gave him; he has filled gaps in my knowledge of army procedures and other practical matters. Always he has taken my project seriously and helped it on its way to completion.

As I see my laboriously typed and retyped pages assume a new identity and dignity on the printed page, I am immensely grateful to Mark Weiman, owner and operator of the Regent Press, for making it happen. Mark approaches the task of bookmaking from the standpoint of an artist. I thank him for bringing his artistic imagination to bear on *Amorette's Watch.*

Prologue
Dream of Amorette's Watch

In my dream I found among my possessions a small white cardboard box with a lid, a box exactly like the one in which, the week before, the jeweler had packed my granddaughter's graduation necklace. In the dream I pulled the lid off with some difficulty, and there inside lay Amorette's watch, circular and flat, turned so that its fluted knob, with the gold loop attached, fitted into a corner. The case was tarnished, but while one half of its surface had darkened to a brownish black, almost obliterating the design engraved upon it, the other half had escaped the damage, whatever it was, and had retained its mellow old gold character. Thus at least half of the little church was still clearly visible, and that was how I knew for certain that this was Amorette's watch, inexplicably returned to me.

I never for a moment believed that I had had it all the time, that I had simply mislaid it and then, twelve years later, had found it again. For even in my dream I remembered that I had seen the small empty box on my dresser on that Fourth of July, 1988, though it had been a different box then, not cardboard but metal of some kind, covered with gold plush and lined with ivory satin. My dreaming self retained the knowledge that the watch had really been gone for twelve years, yet I knew with equal certainty that it was there in front of me.

I wanted to clean the gold case, to polish away the discoloration. Especially I wanted the little church on the hill, with its low tower and its simple doorway, to be whole and beautiful again. A jeweler stepped forward and suggested I try rubbing the watch with cooking oil to lighten the dark areas. He told me to break off some small twigs from a shrub that conveniently appeared and dig away at the dirt and tarnish that had accumulated in the monogram and in the incised lines of the church's walls and roof. I didn't want to follow his advice; I thought the oil would be useless and the twigs would deface those flowing letters, A E F, and that lovely, innocent shrine.

Irresolute, wondering where it had been to have suffered such neglect, I stood with the damaged watch in my hand until I woke up.

For a long time I was sad and angry about the theft of Amorette's watch—sad of course because it was gone and angry because my husband and I thought we knew who had taken it. We thought it was Cap's thirty-five-year-old son, who was addicted to heroin. Up until then Jon had never stolen any of my things, but he had stolen from his father repeatedly. In the days before the theft Jon and his friend Patrick had been working in our bedroom, replacing the ropes in the double-hung windows and painting the window frames. It would have been easy for them to see and examine the jewelry that I kept on my dresser.

On the morning of the Fourth, Cap set out for Marin County by himself. Jon had invited him to a barbecue given by some of his friends. I had made a plan with a woman friend to accompany her and her dog on a walk along the shore of the bay. As I was getting ready, Jon called. I couldn't tell exactly what he wanted.

"Have you seen Cap?" I asked.

"No I haven't. I wondered if you knew where he was."

"He left here about an hour ago. Maybe he's having trouble finding the place." I assumed Jon was calling from his friends' house in Marin.

"Well I haven't seen him," Jon said.

"I'm going for a walk now," I told him, "but you can leave a message on the machine if you have any news. Or you can ask him to."

When I came home from my walk there was no message on the machine from either of them. I fixed myself some supper and then went upstairs, thinking I would start packing for our upcoming trip to Yosemite. That was when I saw the empty boxes on my dresser. Most of my necklaces were gone—the pearls Cap and I selected together in Hong Kong, the piece of green and white jade, suspended on a slender chain, that we found in a shop on Union Square, the amethyst Cap brought home one day from Chinatown. The small gold box was empty too.

Shaken with anger and disbelief, I called the police. An officer arrived promptly. He had just finished his report and left when Cap came in the front door; they must have passed each other in the yard. I told him what had happened.

"It was Jon," he said, his voice rough with the painful certainty. "It must have been Jon."

"The house was empty all afternoon," I said.

He thought for a moment. "I wondered where he was. I

had a hell of a time finding that place in Marin, took me two hours, and then when I did find it he wasn't there. He didn't show up till dinnertime."

"But I locked the house before I left," I said.

"Oh he could get in all right. He knows how to get in."

The police officer who came out the next morning could give us little hope of recovering the jewelry even if we stayed in town, so we went to Yosemite as planned. When we returned there were three messages from Jon on the answering machine, each with a separate piece of information: "I know where your jewelry is" . . . . "I've been talking to the police" . . . . "There's a pawnshop in San Rafael. . . ." The hope that Amorette's watch would be there sent me across the bay before I unpacked my bags.

I found the pawnshop easily. The dealer showed me all the watches that were anything like the one I described, but Amorette's was not among them. He knew Jon, and Patrick too; he did business with them from time to time. He assured me, however, that he himself did not deal in stolen goods.

With no proof I nonetheless blamed Jon for the loss of Amorette's watch. Not the watch as a piece of jewelry but the watch as a tangible reminder of the existence of Amorette E. Foote. That was its main value and always had been. After the burglary an image of Amorette's watch lying forlornly in some pawnbroker's glass case, barely distinguishable from the other old watches crowding in upon it, kept floating up to the surface of my mind. I could not claim that her timepiece was unique, yet the little church on the lid was, to my mind, hers alone, since it symbolized a part of her life, the religious part, the years of Christian service that I knew she had given.

I first learned about her religious work from my mother when I was a child. She said Amorette had been a Methodist deaconess. Deaconesses, she told me, were supported by the church; they did good works and lived together in institutions. I thought it sounded unpleasant to live in an institution. My uncle was the superintendent of an institution for "delinquent" children, as they were then called. I had visited there; my uncle and aunt and cousin lived in the building. It had long bare halls and rubber treads on the stairs—spongy gray-brown mats, held in place by metal edges. I knew I would not want to live in a place like that.

When Mother was explaining Amorette's work to me, she showed me a small photograph, enclosed in an oval frame. I saw an unsmiling older woman who was looking to the side, away from me. She wore a little black bonnet that perched high on her head and tied in a bow under her chin. The picture was faded and the woman looked faded too. "This is Daddy's mother," my mother told me. She said *is,* but I knew Amorette was dead. I did not think of the woman in the picture as my grandmother. She was just my father's mother.

Off and on in my childhood years Mother told me some other things about Amorette. Her maiden name was Rich. She married Seth Foot (as he always spelled his surname), who became a soldier and died in the Civil War. She had one child, Dellizon, born in 1860 in rural Ohio. Mother did not know where Amorette got the name Dellizon, which Daddy did not like. She joked that Amorette must have read it in some novel. Where Amorette and Dellizon lived when he was a child and what they lived on Mother did not tell me, perhaps because she did not know. After Daddy graduated from college—at Upper Iowa University in the town of Fayette—he and Amorette "went to South Dakota and tried to homestead," Mother said. They did not make a success of this venture, she thought, and they did not stay long. Eventually Amorette became a deaconess.

Some time in her life she acquired a rocking chair. This fact I knew from observation, for we had it in our house when I was young. I have it now. It is armless, made of oak, stained a reddish brown. The back is one piece of wood, solid in the upper part and cut into eight narrow, almost Gothic arches in the lower half. The upper part curves around the shoulders; the lower part, despite the cutout portions, reassuringly supports the small of the back. Everything about the chair is comfortable, down to the familiar creaking sound it makes when it is rocked. I always found it a perfect chair to nurse a baby in.

Out of the objects that had belonged to Amorette and the stories I was told, I fashioned my own picture of her: a sober-faced old lady, with her watch in her pocket and her bonnet on her head, sitting alone in her room in an institution. She sits quietly in her chair, not even rocking. She is very far away from me.

The watch was never found. The dream of its miraculous return did not come true. Yet the loss of this memento of Amorette's life also conferred a gain, for it awakened me, not without some prodding, to the need to know the substance of the life itself. The prodding came from Henry, my present husband, who has done some biographical writing himself. At a family gathering I had announced that I was writing an essay about my grandmother, "though I really don't know anything about her." I had in mind some kind of impressionistic piece about a vanished era.

"How can you write about her if you 'don't know anything about her'?" Henry asked. He was right of course. I began to search, looking first in the book of Foote family genealogy, which had been my father's. The dates I found there told me that Amorette and I were not so distant from each other after all. In one respect at least, we were alike: we both came of age in a time of war.

Amorette married at the age of twenty, in 1859. That was the year when John Brown led his ill-fated raid on Harper's Ferry and was hanged for what some saw as treason, others as heroic martyrdom. On April 14, 1860, Amorette gave birth to a son. One year later, to the day, the American garrison surrendered to the Confederates at Fort Sumter, and the four-year conflict, deadliest in our history, officially began.

I entered college in the September before Pearl Harbor. The year was 1941 and I was seventeen. I had my twenty-first birthday two days before V-E Day, May 7, 1945, the day Germany surrendered to the Allies. A month later I graduated; two weeks after that I married David, my college sweetheart. On V-J Day, August 13, the day the war with Japan ended, I heard the newsboys crying EXTRA as I came out the door of a medical office building on Wilshire Boulevard. A few minutes earlier my doctor had confirmed my belief that I was pregnant. Both pieces of news were equally joyful to me.

We two young women, then, came into adulthood during wars of profound consequence to our nation's history. Both of us—Amorette more than I, but I to some degree—felt the effect of the war on our intimate personal lives. Both of us married young; both of us bore our sons six days short of nine months after the wedding day. As I grasp the fact of this elemental kinship I know myself to be Amorette's granddaughter. Reaching across one whole generation—in my case a very long generation—I take her hand.

Part One:

The Early Years

Chapter 1
Amorette

Amorette was born to John and Polly Rich on July 3, 1839, in Lenox Township, Madison County, New York. The northern boundary of their town was the beautiful Oneida Lake, abundant in fish and bordered by luxuriant woods. The site where Lenox stood had for many years been the center of the region known as the Oneida lands, home and hunting ground of the Oneida Indians. But Lenox acquired a new fame in the second decade of the nineteenth century, when a former Indian village within its borders was chosen by the State of New York to be a major construction site for the Erie Canal. It may have been the prospect of work on the huge project that brought John Rich to Lenox from his native Vermont.

In 1824, when he was twenty-five, John married Polly Gilbert. That is the way her name is written in the Rich Family Record, which appears to be a page from their family Bible. But Polly was only a nickname; her given name was Mary Stuart Gilbert. The reason I know this is that I was almost named for her. My father wanted to give me his grandmother's name, but he happened to mention it to his eldest daughter, who was pregnant at the same time my mother was. The other baby was born first, it was a girl, and she got the name. I have never been dissatisfied with the name I was given, and it touches me to think of my father's evident affection for his grandmother.

Amorette had a sister, Adelia Dorleska, nearly fourteen years older than herself, and two brothers, Alison Newton and Angello Everette. There had been another boy, but he died young, and Adelia was the only one of the children who could have known him. After Angello was born, Polly had no more children for six years. Then on the day before the Fourth of July she gave birth to Amorette Elizabeth. The person who inscribed her birth in the Rich Family Record, probably John Rich, wrote her name as "Amoret Eliza," and a census-taker, eleven years later, gave it an Italian flavor by spelling it "Amaretta." But Amorette herself always preferred the Gallic spelling, though she probably pronounced it (or at least my parents pronounced it) in quite an American way—Am' o ret.

What prompted John and Polly to give their children these ornate, exotic names? They are alliterative, for one thing. Not only do all of them begin with A, three of them share the initials A. E.. One name seems French, though not quite. The story in the family was that Amorette was the French word for "little love," but in fact the French word for love is *amour,* not *amor,* and the French word *amourette* means "love affair. . . a passing fancy," surely not what Amorette's parents had in mind when they named their daughter. Angello's name sounds Italian, but the Italian *Angelo* contains only one *l. Dorleska* looks Polish or Russian to me, but someone who knows these languages might find fault with it too. The last three names suggest to me that John or Polly or both of them might have been recalling, by ear, the names of ancestors who came from these European countries. The possibility that a Continental strain lightened the heavily Anglo-Saxon blood of my family intrigues and pleases me.

During the decade that ended with Amorette's birth, the Canal was carrying a steady stream of goods toward eastern markets. But the Canal also led westward, and it beckoned to John Rich. When Amorette was four or five years old she had to leave her home by the lake and make a long journey. The first leg of their trip took them across the state of New York, and the logical way to travel, for this family, would have been the Canal.

To a four-year-old girl the sleeping arrangements on a canal boat must have seemed strange indeed. After the evening meal was eaten (in the big cabin below the deck) she would watch some men pull long narrow shelves out from the walls. Then a curtain would be drawn across the center of the cabin, dividing it into two sleeping spaces, one for the women and girls, the other for the men and boys. Mary Rich would need a shelf for herself, but Adelia would probably share hers with Amorette. That way, if the boat accidentally bumped into the bank of the canal in the night, as sometimes happened, she could hold Amorette fast in the crook of her arm.

The summer I was four I took a long journey west with my father, traveling by train from Omaha to California. We were not moving to a new home, however; we were just paying a visit. I remember sleeping on the train—the porter pulled down a piece of the wall and it turned into a bed. I had to climb a ladder to get into it.

Daddy had taken me on this trip because he wanted to

introduce me to the son and daughters who had never seen me. I remember Daddy's daughter Marjorie the best because she was the mother of the girl who, a few weeks older than I, was my niece. It seemed funny that I could be Mary Stuart's aunt. She and I were given dresses just alike—yellow with smocking at the yoke and a white ruffled collar. We had our photograph taken, sitting side by side, looking like twins.

I must have met Daddy's other two children, and their children, but I don't remember them, and my memories of the train ride back to Omaha are so mingled with the anecdotes my father later told my mother that I can't tell one from the other. But my memory of our return to our house is entirely my own. I am standing on the big screened porch, where you could look down the hill and across the tops of the houses and the trees until you could see silvery patches of river and then the bluffs on the Iowa side. I always liked standing there and having Mama and Daddy point out these things to me. But on this evening I don't look beyond our screened porch. I look at the table, at the plates and glasses and knives and forks and spoons, and at Mama bringing out the food. I am entirely happy.

Nobody ever explained to me why my mother did not go with us on that trip. A letter of hers that I recently found shows that she kept Daddy's office open in his absence, collecting bill payments, making appointments and talking with patients who just dropped in to say hello, but surely someone else could have done that. Probably my parents decided that introducing his new child to his other family was something my father needed to do alone.

And so the two of us had our summer adventure together, and the three of us had our loving reunion in the home which has sometimes seemed, in retrospect, to be part of an idyll, a lost Arcadia. The summer I was four was the last summer we spent there. Daddy had a stroke in the autumn; he was bedridden for months and partially paralyzed thereafter, unable ever to work again. There were good, practical reasons why we should move away from the house on the hill, with its half-acre of land, with its lawns and gardens and orchard and—most compelling reason of all—with the steep street that was the only way to get up there. One of the first things Mother had to do that winter was learn to drive, and the snowy, icy hill was a terror to her.

The Riches probably had good, practical reasons for moving away from Lenox Township, too. Still, I cannot help thinking of Amorette's first home, with its lake and its woods, as her lost Arcadia.

If, as I suppose, the Riches began their journey by way of a canal-boat, they would have disembarked at Buffalo. From there a steamer would have carried them along the lower edge of Lake Erie to a port in northeastern Ohio. This region of the state, known as the Western Reserve, had been settled mainly by New Englanders, which may have been the reason John Rich was drawn to it. Their journey ended in Westfield Township, Medina County, about thirty-five miles south of Cleveland. Their neighbors were people much like themselves—northerners, Protestants, white. The Indians were gone from Ohio by the time the Riches arrived, and few black people lived in Medina County.

About three years after the move to Ohio John Rich inscribed another marriage in the Family Record. The wedding of Adelia Rich and Otis Holmes was performed by a minister in April of 1847. A picture in Mary Rich's album—a tiny oval, embossed all around in a pattern of leaves—shows a young woman with long dark ringlets and calm, serious eyes. Underneath it someone has penciled "Ada Holmes." A little over a year after the wedding Adelia and Otis had a daughter, whom they called Dorleska. The little girl was born on May 27, 1848, and at first the mother and baby thrived. But in the hot days of late summer—the season for cholera and other infectious diseases—Dorleska fell ill. Nothing that Adelia or anyone else did for her could save the child. "Dorleska Elnora Holmes died August 11 1848," the Family Record reads. Before the month was out another entry had to be made: "Adelia Dorleska Holmes died August 28th 1848." Seventeen days separated the baby's death from that of her mother. Adelia's infant daughter had lived two and a half months; Adelia herself, twenty-one years.

Amorette was nine when this grief came upon the family. Suddenly, in one summer month, she lost the sister who had always been a presence in her life and the tiny niece who had just entered it. Like so many children of her time, Amorette was given an early lesson in bereavement.

In the census records for 1850 I found that John Rich, aged fifty-one, was a carriage-maker by trade. Evidently he had taught some of the skills of his craft to his sons. Newton, now nineteen, called himself a carpenter and joiner; seventeen-year-old Angello said he was a wagon maker. No occupation is given for John's wife, who has dropped the nickname of Polly in favor of her Christian name, Mary. Forty-nine years old now, Mary is not described as "housewife," nor are any of the other women on this page of the census record, though we can guess what their occupation must have been. In fact, the census form itself excludes the possibility that a woman might have an occupation, calling only for the "Profession, Occupation, or Trade of each Male Person over 15 years of age."

Newton at nineteen may be a full-time craftsman now, but

he is not long away from his school days; both he and Angello attended school during the preceding year, as did eleven-year-old Amorette. From then on she would spend more time in a schoolroom, year by year, than her brothers had been required to do, for Ohio had just passed laws extending the school year to at least thirty-six weeks.

In the nation's history, the year 1850 is remembered most of all for the infamous Fugitive Slave Law, which required all citizens, Northern as well as Southern, to cooperate in returning runaway slaves to their owners. This new policy of the federal government was violently opposed by many Northerners. Among Ohioans the anti-slavery sentiment was probably strongest in the Western Reserve, though it was felt in many parts of the state. Indeed, in the decade before the war Ohio contained more stations of the Underground Railroad than any other state. Where the Rich family stood on the slavery question is unknown to me, but my guess, based on geography and family tradition, is that they opposed the Fugitive Slave Law but did not consider themselves abolitionists.

In 1854, the year when the Kansas-Nebraska Act widened the gulf between slave-owners and opponents of slavery, the Riches suffered another kind of rupture: on the last day of September John Rich died. His death came little more than a month after the marriage of his elder son, Newton. These two events, it appears, altered the circumstances of Mary's household. Now, instead of three breadwinners, there was only one, twenty-one-year-old Angello. Yet Mary was not such a helpless widow as her circumstances suggest; in fact, she seems to have been something of a businesswoman. After John's death she bought a ten-acre piece of land, bringing her total holdings to twenty-one acres.

It is hard to know what she did with the land, though,

once both of her sons moved away. In 1857 they migrated westward, their destination Iowa. This time Mary stayed behind, but Amorette went west with her brothers. It surprises me that Mary would have allowed her daughter, who would not be eighteen until July, to accompany Newton and his wife and Angello to a distant and for them unfamiliar region. I had believed that girls in nineteenth-century America were protected at home until their marriage, yet here is this seventeen-year-old girl, going hundreds of miles away from her mother's watchful care. I have to wonder what incentive was so powerful as to impel Mary to give her consent.

For the two brothers, the incentive was probably economic, first of all. Probably they were caught up in the enthusiasm, widespread at the time, for the "beautiful land," as the Indian name for the state was sometimes translated. Iowa land *was* beautiful, its softly contoured prairies pleasing to the eye and yielding to the plow, and the price of land was low, at first. In the more settled states to the east, all through the 1850s, a chorus of voices had been urging residents to pack up and go to Iowa. Most vociferous were the railroads, which by the mid-fifties had opened three lines as far as the east bank of the Mississippi. Naturally, guidebooks in large numbers made their appearance. One of them, *Iowa As It Is,* personified the new state as a generous female figure, almost a new Ceres. Her resources were "inexhaustible"; she claimed the attention of "every class and sex of the energetic, the industrious and the ambitious," an appeal clearly directed not only to men but to their wives and sisters as well.

It may have been the desire to share her brothers' adventure that drew Amorette to Iowa. Perhaps she was taken along to keep house for Angello, who had no wife as yet. Perhaps in Amorette's vision of the beautiful land lay the hope that she would find love there, meaning, for a girl like her, that she would find a husband. Mary Rich may have given her silent blessing to such a quest.

I am going to assume that the Riches made their journey by covered wagon. At this date it would have been possible to go all the way to Iowa "on the cars," as people said then, for in 1856 the Rock Island had finished building a spindly railroad bridge across the river at Davenport. But travel with a wagon and a team was cheap, especially for young men who knew how to make and repair wagons, and, compared to the space allotted to railroad passengers, a covered wagon was capacious. It would hold the tools and quilts and dishes and pots and pans that would be needed as soon as the party reached its destination—Clayton County, in the northeastern part of the state, bounded on its eastern edge by the Mississippi River.

The young travelers would have to cross two-thirds of Ohio and all of Indiana, then angle northwest through Illinois and Wisconsin before they came to the Mississippi. And when they did reach its broad expanse (for it was a big river even this far north) they would have to stand in line, possibly for days, to put their wagon and team on the ferry. Or perhaps it was two wagons—one driven by Newton, the other by Angello—that stood in the line at Prairie du Chien. Once aboard, they would watch the level Wisconsin floodplain recede behind them as they neared the soaring bluffs on the Iowa side. One of the bluffs, it was said, was the highest point along the entire course of the river.

When the ferry pulled in alongside the boat landing at McGregor, a rope would be tossed to a man at the river's edge, who would snake it around a stump. Then the planks would be let down and a crowd of men, women, children, draft animals and wagons would stream off the boat at the foot of the town's main street, which ran up a gulch so narrow that its sides rose directly behind the frame and brick buildings that lined it. Newly arrived emigrants who wished to leave McGregor for the more level land to the west would first have to drive to the head of the street and then climb a tortuous road up a steep, forested ravine. At the top of the road, where it flattened out, one branch turned southwest toward Farmersburg. I think that might have been the direction the Rich party took, because some time in 1857 Angello made the acquaintance of a young man named Seth Foot, whose father owned a farm a few miles from the little town.

Notwithstanding the claims of the boosters, 1857 turned out to be an inopportune year in which to emigrate to Iowa. As early as January hard times had begun to be felt in Clayton County. In the years just past, ambitious farmers had been buying more land, at higher prices, than they could afford; now they were having difficulty meeting their mortgage payments. The precarious state of the banks—not only in Iowa but across the country—contributed to the widespread mood of anxiety. Then, at the end of the summer, the Panic of 1857 struck the nation. The closure of one bank in New York in late August led to a wave of bank closures. Factories shut down, businesses closed their doors, workers lost their jobs. In farm states like Iowa, prices for crops fell precipitously. For most of the country the depression that followed the Panic was short-lived, but in the Upper Mississippi Valley two years passed with little sign of relief. The price of grain was so low that a farmer saw no point in planting a crop for the market. For some communities, barter was the favored means of exchange.

In this inauspicious time and place—that is, in the summer of 1859 in a town on the Upper Mississippi—Amorette married Seth Foot. They were married on July 20 in Prairie du Chien, Wisconsin, by a justice of the peace. The facts in the story make the marriage sound like an elopement. Why did Seth and Amorette go to another state to be married? It was not that their marriage would have been illegal in Iowa, for Amorette, at twenty, had reached the age of consent and Seth was twenty-five that year. And why choose a justice of the peace instead of being married by a minister, as Adelia had been? My best guess is that Seth and Amorette wanted to marry quickly and quietly, telling as few people as possible. But why?

Chapter 2
Seth

Seth's picture hung in our front hall when I was a child. "Grandfather Foote's portrait," as we called it, showed a serious-looking soldier, head and shoulders only, in a dark blue uniform. The soldier's hair was reddish-brown, as was his short beard. The beard, indeed, was much like the goatee that my father wore, and about the same color. Grandfather's cheeks were pink, a little too pink, I always thought. Nevertheless, I was awed by his air of military authority. I had been told that he was an officer in the Union army, and I had it in my mind that he was at least a captain, perhaps a major.

Daddy still possessed his father's shoulder straps, the same ones he was wearing in the portrait. I could hold them in my hand, and did, feeling surprised at their weight. Curved to fit a man's shoulder, they were stiff, metallic, tarnished. I thought they must have been uncomfortable to wear. I had no way of knowing, at the time, that what I was holding in my hand were the shoulder straps of a second lieutenant.

I did not hear any stories about Grandfather Foote's war experience, only a matter-of-fact explanation, from my mother, of the reason for his death. "He didn't die of wounds received in battle," she said. "He died of disease." She spoke as if this was an important distinction. She did not identify the disease, which probably meant that she had never been told its name, for when she told stories about diseases, such as the typhoid fever epidemic that struck Omaha while she was in nurses' training, she almost always named them.

The letters Seth wrote when he was in the army had, like the portrait, been handed down from Amorette to my father and after his death to my mother. I knew of their existence—we called them "the Civil War letters"—but I never read them, or even handled them, when I was a child. It was not until I was packing up Mother's things to move her out of her house (she was eighty-three by this time) that I first saw them. I scanned them briefly, having more immediate concerns on my mind, and put them back into the box where I had found them. It went into a rented storage space, along with most of Mother's possessions, when I moved her to Berkeley to be closer to me.

A while later, realizing that Mother would never use those possessions again—never read her books or knit her potholders or type her letters to City Hall—I moved them out of storage and into the basement of our house. There I intended to sort them and make some kind of disposition of them. I called a rare book dealer to see if I could sell any of the books; he came out and selected one or two. I gave household goods to family members, keeping some for myself, and of course I threw some things away. And near the end of the procedure I looked for the Civil War letters and they were gone. I blamed myself bitterly for not giving the letters the care they deserved. All I remembered from my quick glance at them was that they were addressed to Amorette and that some of them were mailed from Fort Laramie. It had come as a surprise, I recalled, to think that a Civil War soldier would be stationed in Wyoming.

It was in the summer of 1979 that I lost Seth's letters. Seventeen years later, when I began to write Amorette's story and realized that it must include Seth's, I felt their loss even more keenly. Without them I had either to give up my effort to know my grandfather, which I did not want to do, or to look for other documentary evidence. So I began to look, and in libraries, in a county courthouse, in the nation's capitol and in my own house, I discovered more traces of Seth's life than I ever expected to find.

Seth was born in northeastern Ohio on January 25, 1834, the oldest son of Moses and Maria Foot. (Unlike most members of the Foote clan, Moses had dropped the vestigial *e* in the surname, and Seth would follow his example.) Seth's father, like Amorette's father John Rich, was a New Englander by birth. Moses Foot was born in 1812 in Connecticut, home of his ancestors for six generations. He moved with his parents to New York State when he was five. It was the beginning of a westward movement, for this family, that roughly paralleled the movements of the Rich family and countless others in those stirring first decades of the new century of independence.

It was probably in New York State that Moses found his religious faith. The writer who, after Moses' death, composed his obituary, reported that "he embraced religion in his early youth" and remained "firm in that faith" throughout his life. The years of his early youth coincided with the time of the Second Great Awakening, an evangelical Protestant movement especially strong in upstate New York. It is likely that as a boy

Moses was influenced by the singing, shouting, praying revival services that characterized the movement. Before he was fully grown he and his family went west again. They settled this time in Huron County, Ohio, just two counties west of Medina County, where the Rich family would make their home. At the age of twenty-one Moses married Maria Palmer. She bore four more children after Seth—three boys and a girl. Then, still in her thirties, she died. Seth was fourteen when he lost his mother; David, his youngest brother, was only five.

Maintaining a home for such a family must have difficult for the widower; clearly, Moses needed a wife. A little over a year after Maria's death he married Mrs. Mary Lockwood Lester, who brought her three children into his household. Now there were ten people—two adults and eight children—living under Moses' roof. The 1850 census form for Fitchville, Huron County, Ohio, lists them all by name: first Moses and Mary, then Seth, Henry, Hamilton, Tama and David Foot; and after them Nelson, Lois and Gurdon Lester. The spelling of the last name on the list is unusual. Perhaps it was the way someone pronounced *Gordon*.

The census form describes Moses Foot as "Farmer." Sixteen-year-old Seth Foot and eighteen-year-old Nelson Lester were also calling themselves farmers. A bit of adolescent boasting was operating here, I think; the likelihood is that Seth and Nelson helped Moses with the work of his farm. The two of them had also been schoolboys in the year just past. That fact alone tells me that Seth had at least some high school education.

In 1855 Moses purchased a piece of land in Clayton County, Iowa. Seth would be the first in the family to make the move, but he did not go immediately. A year after his father bought the land Seth, twenty-two years old now, was still in Ohio. That was the year he started borrowing money. In April he borrowed forty dollars from a man named Philander Austin, a sum that sounds negligible today but was worth about nine hundred of today's dollars. Seth did not repay the loan, but Austin evidently trusted him enough to go into business with him, on credit, the next year. On Valentine's Day, 1857, Seth and Austin, acting together, incurred three debts. Here is how I have reconstructed the events of that day:

The two men go first to the firm of D. D. Van Vretten & Company and buy a quantity of boots and shoes, which they intend to sell. The bill comes to just over $550. Next they go to a man named David Davis and borrow $125 from him, with a promise to repay him on the first of September. Before the day is over they borrow $110 from someone named James Probert. In each instance both Austin and Seth sign their names to the notes.

At the end of June Seth added to his personal indebtedness by borrowing another forty dollars from Austin, promising to repay the sum on "the first of December next with ten per cent interest." Neither Seth nor Austin, nor almost anyone else, could have foreseen what would happen in the world of finance that summer. On the day the Davis loan came due, the Panic of 1857 was sweeping over the country. As September gave way to October, ordinary people were coming to the stunned realization that there seemed to be no money. Under such circumstances, "Woe betide the man who was in debt," as an Iowa historian put it. Seth, however, went more deeply into debt that month. On October 20 he and his brother Henry gave a note, promising to pay fifty dollars within three months, to Moses' brother Walter Foot.

All these transactions — the initial purchase from the Van Vretten Company and the succession of loans — took place in and around Seth's home town of Fitchville, but within a year he moved to Clayton County. Apparently he took with him a load of boots and shoes, and possibly some horse harnesses, hoping to sell them. Economic conditions in northeastern Iowa were no better than they had been in Ohio—if anything they were worse—but at least Seth had put some distance between himself and his creditors. Back in Fitchville, the loans were coming due. In April of 1858 Austin made a partial payment on Davis's loan, now seven months overdue. That satisfied Davis for a time. The Van Vretten Company, however, grew anxious and took its complaint to court. A judgment was entered against Seth and Austin; the debt now came to $558. Austin paid this amount in full, including Seth's share.

Fall and winter passed, and Seth did not reimburse Austin. In the spring of 1859 he incurred another debt, a small one. He was doing a little farming, probably on Moses' land. He owned a cow and a horse, but he did not own any oxen, and he did not have adequate pasture for the cow and the horse. A neighbor, A. G. Cloudman, pastured Seth's cow for ten weeks and four days that spring, and his horse for forty-three days. Cloudman also rented Seth a pair of oxen to work for one and a half days. His total bill was $8.76, which Seth did not pay.

Some time that spring David Davis concluded that he had been patient long enough. He went to court, and on June 1 the Justice of the Peace entered another judgment against Seth and Austin. Austin paid this debt and also repaid, in full, the loan from James Probert. With these payments Austin cleared the debts he and Seth had incurred on that February day in 1857 and freed Seth and himself from further legal action in Huron County. Quite naturally, he wanted Seth to contribute his share of the repayment, but Seth did not do so.

Seven weeks later, in Prairie du Chien, Wisconsin, Seth married Amorette. The foregoing review of his unhappy financial history suggests a reason why the young couple crossed to the other side of the Mississippi for their marriage: perhaps Seth was trying to prevent his creditors from knowing where he was. Perhaps he hoped that if his name were not written into the record in Clayton County, they would be unable to find him. By this time, however, Seth's chief creditor was Austin, and he did know where Seth was.

The facts and figures I have reported above were contained in a packet sent to me from Elkader, Iowa. I had asked the Clayton County Recorder to mail me a copy of Seth's will. What I received, in addition, was a sheaf of copies of the documents supporting the settlement of his estate, twenty legal-sized pages in all. I squinted over them, studied them and did my best to construct a narrative from them. They were extremely useful; still, I had a desire to see and touch the originals. And so, on a trip to northeastern Iowa to visit the places where Seth and Amorette had lived, I stopped in at the county courthouse in Elkader.

The sun was fast dispelling an early fog as I turned off the highway and drove downhill to the main part of town. The courthouse, I found, was a stately old building of rose-colored brick, graced with a tall white cupola. Once inside, I presented myself at the counter of the Recorder's Office and asked to see the file pertaining to the settlement of Seth Foot's will. The three women in the office remembered my name, and one recalled making the copies and mailing them to me. At first they could not quite see why I needed to look at the originals when I had all those copies. Getting the file would require someone to go high up into the top story, they said. But someone did, and after a few minutes she brought me two old envelopes. Opening them, she pulled out two stacks of closely folded sheets of paper and laid them before me.

As I stood at the counter, I began to unfold them and look eagerly through them. I soon realized, however, that I needed to spread them out and study them, one by one, and I asked Jean Welsh, the Recorder, if there was a place where I could sit down. She looked around the room and said I could sit at Susan's desk. I didn't know if Susan was absent that day or if she was one of the three and I had dispossessed her. But given permission, I sat down and lost myself in the strange, sad story of my grandfather's financial affairs.

Almost all the documents were familiar to me from the copies, but the originals had an authenticity that no photocopy could possess. Where the copyist had laid two or three small papers on the glass at once, I now saw each one separately. I saw the colors of paper (blue or brownish-white) and of ink (blue-black and brown). I came upon Seth's signature; I saw the letters his hand had formed, traced in the ink into which he had dipped his pen. The sight assured me, in a new and tangible way, of his existence. For a moment I felt the presence of the living man.

I kept turning over the papers, recognizing the contents of each one, until I had almost reached the bottom of the pile. There I discovered a document so small that it had escaped the notice of the copyist. Signed individually by Seth and his brother Henry, it was the promise made to their Uncle Walter. It was written on a piece of paper less than three inches from top to bottom, and it read as follows:

$50. For value received I promise to pay Walter Foot or bearer fifty dollars three months from date with use.

Fitchville
Oct the 20th, 1857.
Seth Foot
Henry R. Foot

I held this scrap of paper in my hand, feeling the pathos of it. Never repaid in Seth's lifetime, it must have brought disappointment to his uncle, embarrassment to his father. Yet I do not mean to censure Seth. More than anything, I feel sorry for him. He and Austin overextended themselves financially at a time when their country was overextending itself. Somehow Austin managed to meet their joint obligations, while Seth did not. I will not attempt to guess why, but I have no doubt that this small pile of debts laid a heavy burden on the marriage of Seth and Amorette.

A friend who read this part of my grandmother's story asked me incredulously, "Why did she marry him? Did she know he owed all that money?" While I cannot answer the second question, I will venture an opinion on the first. I think Amorette was in love with Seth, and that even if she did know about the debts she did not let her knowledge prevent her from marrying him.

My friend asked one more question: "How did Amorette's mother feel about her daughter's marriage?" On this point I can be more sure. Considering that Mary Rich had never met Seth, I think she would have been dubious about his prospects and anxious about Amorette's future. Most mothers would have felt that way. Within a few months Mary would have even greater reason for concern.

Chapter 3
Husband and Wife

I think that Seth and Amorette made their first home on the land that Moses owned in Wagner Township. Whether there was already a house on it when Moses bought it, or whether Seth built one, I do not know. But if we envision the two of them starting their married life in a small frame house, or cabin, on rising ground near the edge of a grove of trees, it will give us a setting for the actions that almost certainly did take place during the first eight months of their marriage.

The trees would have offered some protection from the extremes of the prairie climate, tempering the summer heat, diminishing the force of the winter winds. From the economic climate in that third summer of depression, there was less protection. We can see what its impact upon the young couple must have been if we look at an issue of the *Clayton County Journal,* published a week and a day after they were married. On one page are two notices headed "Sheriff's Sales of Real

Estate"—bleak evidence of the prevailing hard times. The first announces that a farm belonging to a man named Talbot will be sold to the highest bidder in a little more than a week's time. The sheriff has taken the property, the notice says, to satisfy a claim for $323. The second announces the impending sale of a piece of village property, formerly owned by a husband and wife but now taken by the sheriff to satisfy a claim for $286. Either of these debts was considerably smaller than Seth's debt to Philander Austin. On the same page of the *Journal* two boot-and-shoe merchants are advertising their wares. One offers to sell at the "LOWEST POSSIBLE RATES"; the other vows that he will not be undersold. This was the sort of competition Seth's fledgling boot-and-shoe business faced in Clayton County.

On another page, however, the *Journal* holds out hope of rescue from the hard economic times: "the news from the mines at Cherry Creek continues to be of the most cheering character, new and rich discoveries are being made every day, and the country is fast filling up with adventurers in search of the shining dust." Cherry Creek, which runs through what is now downtown Denver, was the site of early discoveries in the so-called Pike's Peak Gold Rush. The use of the Peak's name was always a misnomer; no gold of any consequence was found on the mountain itself. The most productive finds were made on Cherry Creek, Clear Creek and the headwaters of the South Platte—all of them at least sixty miles to the north. But the appeal of the short, alliterative name was irresistible to headline writers and goldseekers alike and it soon became almost universally accepted.

As summer moved into fall, the *Clayton County Journal* grew more cautious about the prospects for gold. On August 25 the editor says that the news from the mines is discouraging, and on September 22, under the heading of "Pike's Peak

News," he concludes that "no reliance can be placed" on the reports that are coming in from the Rocky Mountains. If Seth had already been considering a trip to the mines, this latest report would have given him something to think about. Amorette, however, had something else to think about in those last days of September. By that time she was probably quite certain that she was pregnant.

If she was like most young women of her time, she was frightened at the prospect. The link in women's minds between childbirth and death was one of long standing. In the middle of the nineteenth century a woman named Georgina Kirby wrote in her journal, "'I think that perhaps I may die and my baby live.'" I have seen the same fear expressed in poems by English women of the eighteenth century. It was strongest among women who were pregnant for the first time—and with good reason. Judith Walzer Leavitt, the historian who quoted Kirby, reports that the maternal death rate was highest for women who were bearing their first child. This fact could not have gone unnoticed by women of Amorette's generation. She herself surely remembered her sister Adelia, dead within three months of her first baby's birth. Though childbirth may not have directly caused Adelia's death, the association was there. Nevertheless I hope Amorette also felt joy at the prospect of becoming a mother, and the story of the rest of her life tells me that she did.

By the end of October the *Clayton County Journal* had changed its mind about the veracity of the reports from the Rocky Mountains. Under the heading "THE GOLD NEWS" the editor writes, "The accounts from the gold mines are becoming more flattering, and it appears there is some reality in the gold tales." Three weeks later he is even more optimistic. Letters received from Denver now cause him to believe that the Rocky

Mountain mines are as rich in gold as those in California.

I cannot tell exactly when Seth conceived the idea of going to the mines, but an event that took place in early November may have urged him in that direction. The event was the arrival in Clayton County of Philander Austin. The trip from Ohio cost Austin forty dollars, but he evidently believed the expense would be justified if he could collect the money Seth owed him. What Austin found when he reached Iowa must have convinced him that his prospects of recovering his money from Seth were not very flattering, as the *Journal* might have said. Before he went back to Ohio, he appointed a local justice of the peace, Asaph Tuttle, to act as his collection agent.

Seth did not have to look far to find companions for an expedition to the mines. Four of the men who signed on—the two Benson brothers, as well as George Wakeman and Franklin Russell—were close neighbors of his. The Benson land adjoined Moses' land on the east, while the Wakeman and Russell properties lay a short distance to the south. Two of the goldseekers were members of Seth's family, his brother Henry and their stepbrother Gurdon Lester. The only comparative outsider in the group was the Reverend W. Mann.

Why would these eight Iowa men, most of them farmers or sons of farmers, decide to put their faith in a dream of extracting gold from a distant mountain? I think the answer was not the same in every case. Gurdon, as subsequent evidence will show, was impelled by a love of adventure and a desire to be where important things were happening. Henry seemed to follow his stepbrother's lead.

The others, however, were probably motivated by the same harsh economic conditions that sent thousands of midwesterners to the goldfields in the lean years from 1857 to 1860.

Though the economy had improved in much of the country, the depression had not relaxed its grip on the Upper Mississippi Valley. In the winter of 1860 Horace Greeley, editor of the influential *New York Weekly Tribune,* made an inspection tour of the region and sent his paper a dismal account of what he was seeing. He placed much of the blame on speculative railroad schemes: "'The West is poor. The collapse of the railroad bubble. . . has spread desolation over the land.'" When he wrote these words, Greeley was in Prairie du Chien, Wisconsin, just across the river from Clayton County.

It seems clear that Seth joined the goldseekers as a way of escaping his immediate money problems. On February 11 Justice Asaph Tuttle was granted the power of attorney that enabled him to transact business for Philander Austin. In March Tuttle was keeping a horse to facilitate his transactions and was charging the upkeep of the horse to Seth's account. On March 14 Tuttle "bought" from Seth the balance of a mortgage, which came to $435. This was not a cash purchase; Seth handed the mortgage paper over to Tuttle, and Tuttle entered a credit for the sum in Seth's account with Austin. (How Seth, in his apparently straitened circumstances, came to be the holder of a mortgage is a mystery I cannot hope to explain.) By this time Seth was undoubtedly making final preparations for the trip, since the starting date was only two and a half weeks away. And just about this time Amorette went home to her mother.

Several things probably brought her to this decision. One was the worrying presence of Austin and Tuttle in her husband's life, acting as a constant reminder of the debts Seth had not paid. Another was the plan for the Pike's Peak expedition, making it clear that Seth would not be with her when the baby was born. Finally, there was the question of financial support for herself and her child, once Seth was gone.

But even if these factors had not existed, Amorette might still have wanted her mother to be with her when she delivered her first child. If she had stayed in Clayton County, she would have had no female relative except her sister-in-law to rely upon when she went into labor, and Newton's wife might not have been available, or suitable. In such circumstances, as historian Leavitt remarks, nineteenth-century women who had previously migrated westward sometimes made extraordinary efforts to return to their mothers' homes for their confinements: "Despite the difficulty of travel during their advanced pregnancies, many women who could not find appropriate women to attend them went home to mother." The phrase itself says a great deal; a young wife went home to mother for comfort, for advice, for reassurance. Mothers had been through childbirth and lived. Their very existence gave their daughters cause to hope.

Amorette probably traveled to her old home by train. A railroad line had been in service from Prairie du Chien to Great Lakes cities for three years, and there was also a train from Cleveland to a point fairly near Mary Rich's home in Westfield Township. The train ride would have been uncomfortable for a woman in the late stage of pregnancy, but it would have been the most efficient way of getting her to her destination.

I imagine Amorette began the journey by traveling the same road she and Seth had taken on their wedding day—across the rolling farm land to the wooded ravine, down the switchback road to the main street of McGregor and along its length to the ferry landing. It is likely that Seth would have driven her this far and would have accompanied her on the ferry ride.

In my mind's eye I see the two of them standing by the railroad track. They are about to go in opposite directions. He will go west into unknown territory, and there is no knowing when he will return. She, impelled by the new life growing inside her, is being drawn east to a bed in her mother's house and a trial from which she may not emerge alive. The husband and wife are both weeping.

PART TWO

New Territories

Chapter 4

Off to the Mines

"G. P. Lester, H. R. Foot, S. Foot, Franklin Russell, Wm. Benson, E. Benson, Geo. Wakeman & Rev. W. Mann started enrout for Pike's Peak Monday, Apr. 2nd 1860 with fifteen head of cattle & three wagons with provisions for six months."

Of course, I thought. S. Foot, that's where he was! The first sentence of the typescript that had just come to me in the mail gave me an answer to a question that had been puzzling me for some time. It was quite early in my research on my grandparents, and I had lost sight of Seth. I knew he married Amorette on July 20, 1859,and that he joined the army on August 28, 1861. But where he was between the two dates I had not been able to discover. The page I held in my hand, part of a typed copy of a diary Gurdon Lester kept in 1860, brought Seth into view.

I learned about the existence of the diary through one of

those lucky chances that sometimes come to researchers. It began in the Bancroft Library of the University of California, that invaluable source of materials on the American West. Having recalled that some of Seth's Civil War letters were mailed from Fort Laramie, I was sitting at a table reading Merrill Mattes's *The Great Platte River Road,* which traces the famous covered wagon road "Via Fort Kearny to Fort Laramie." I was doing this mainly for historical and geographical background, never imagining I would find any of my family in it. But in a chapter on river crossings I came across the name of Gurdon P. Lester. Was that my Gurdon Lester? I thought it probably was; Gurdon was an uncommon name. Mattes gave him only one sentence: "In 1866 stage passenger Gurdon P. Lester says he 'landed 3 miles above Kearney Station,' or about three and a half miles above the fort." That one sentence, however, justified Lester's inclusion in Mattes's bibliography, and it directed me to an article in a 1965 issue of *Nebraska History.* I found the article, discovered that it contained a travel diary Lester had kept in 1866 and—more to the point—learned that he had also kept a travel diary in 1860, when he went "to the Colorado gold fields."

Then the question was, did Seth go with him? I needed to see the 1860 diary. Charles W. Martin, author of the article, reported that the handwritten original was owned by a man named Charles H. D. Smith, who had loaned it to the Joslyn Art Museum of Omaha in 1965 for a special display. But that was more than thirty years ago. Where was the diary now? I wrote a letter to Mr. Smith; it was answered by his widow, who told me that the Joslyn now possessed it, along with a typed copy that her husband had made. I wrote the Joslyn, and someone there made me a copy of the typescript.

The list of names in Gurdon's first sentence and his use of the third person in referring to himself suggest that he is conscious of the historical significance of the record he is beginning. His book will be more than a personal diary, it will tell the story of an expedition. Naturally, Gurdon lists his own name first. Youngest of the three children whom Moses Foot's second wife brought to their marriage, Gurdon Perkins Lester was born in Cayuga County, New York, in 1837. At the time the Pike's Peak journey began, Gurdon was twenty-two years old, Henry Foot was twenty-three and Seth twenty-six.

The three wagons Gurdon mentions in his first sentence no doubt were the customary farm wagons to which a canvas canopy had been added. All that first day the company traveled southward—each wagon pulled by two yoke of oxen, the men walking alongside the wagons, some of them leading cows. At the end of the day they camped on the Turkey River in the town of Elkader; on their second night they camped near Strawberry Point. So far nothing had happened to delay their progress. But "on the 3d day of our travel," as Gurdon reports, "Wakeman accidentally sat the prairie on fire with a prairie grass wad from a shot gun while attempting to shoot a prairie chicken." Seeing the flames, Seth, Gurdon and Russell grabbed whatever they could lay their hands on and began to help Wakeman fight the spreading fire. It took them a full hour to put it out. Meanwhile the rest of the company continued on their way. The firefighters rejoined them at Otter Creek, where they all camped for the night.

Gurdon's word *sat* (above) must mean *set.* I have decided to reproduce his occasional misspellings just as he wrote them because I think they are an appealing part of his character. I see him as a young man who had a zest for writing and an earnest desire to record these new experiences vividly, but who, as a schoolboy, had not paid much attention when spelling lessons were taught.

At this time the men were following a southwesterly course. By Friday, when they reached Vinton, they were about seventy-five miles from home. For part of that distance, it seems, they were being pursued. In this same week Asaph Tuttle, Philander Austin's agent, was driving a horse and buggy to the town of Independence, which lies on the course the men were following. Tuttle charged the expense of the trip to Seth's account with Austin, dating the item April 4. The charge "for self and horse going to Independence & Expences" was eight dollars. But there were damages to pay as well: $18.75 "for triming & reparing Buggy."

It is not clear to me whether Tuttle caught up with Seth that day. I know that he repossessed twenty-two pairs of women's shoes, seven pairs of men's boots and two horse harnesses because he took the cost of them off Seth's account. When and where this transaction took place I cannot tell. But I cherish in my mind a picture of the justice of the peace, urging his horse and buggy over the rutted country roads, in hot pursuit of three wagons, eight men and fifteen head of cattle.

On their first weekend the company began a custom, "spending the sabbath," which they continued, almost without exception, for the rest of the journey. It was a practice recommended by J. Quinn Thornton in a guidebook written for western travelers. "Emigrants," Thornton wrote, "ought to remain in camp on the Sabbath day. Aside from the obligation imposed by the command of God, the cattle will endure more if they are permitted to enjoy this periodical rest." The presence of Reverend Mann in the company made it easy for the Farmersburg men to fulfill their religious obligation. On that first Sunday, while they were camped on a high prairie, he preached twice, first at eleven a.m. and again at five p.m.. He had a smaller congregation at the five o'clock service; Gurdon says that at five p.m. "all the boys left Russell & myself to guard the wagons."

Perhaps Reverend Mann was disappointed by the company's unenthusiastic response to his preaching, or perhaps he and George Wakeman had grown tired of the other men's company. In any case the two of them struck out for themselves before the next sabbath arrived. They promised to rejoin the company at Glenwood, the last town of any consequence before they came to the Missouri River.

On Saturday April 14 the remaining six members camped "about one hundred rods from Middle River." It was an especially pleasant campsite, from the sound of it, well suited for spending the sabbath. "There we found the nicest timber I ever found in Iowa," Gurdon observed. "On this stream there is plenty of Buckeye timber something that I never saw before in the State of Iowa. The Cottonwoods on this stream are the nicest I ever saw." For Gurdon, who seems to be comparing Iowa trees with those he knew in his New York boyhood, April 14 was a day to remember. It was a memorable day for Seth as well, though he did not know it yet. On that day, at her mother's house in Ohio, Amorette gave birth to their son.

For most of that week Gurdon marked off the days with the names of rivers. Leaving the Middle River camp Monday morning, they traveled about twenty-two miles and camped on Middle River again. Tuesday evening found them on Nodaway River, which promised them a tranquil sleep. Wednesday night they were on a high prairie, with no river at hand, but on Thursday night they reached the Nishnabotna. Here, Gurdon reported with obvious pleasure, "we found several of our old acquaintances of Clayton Co." One of them was Amorette's brother Newton—"A.N. Rich," as Gurdon calls him.

On their third sabbath Gurdon, Seth and Henry turned

their cows to grass for the first time. Up until then, evidently, they had been giving the cows feed they had brought in the wagon, but little by little those cows were losing their privileged status: "Before this late date we became tired of leading our cows so we purchased a yoke and put our [cows] in the middle of the team having one yoke forward and one behind which held them perfectly firm we had no difficulty learning our cows to draw." The fact that Gurdon and Seth and Henry had "no difficulty" speaks well for their abilities as cattle-handlers. It was not easy to teach ordinary cows—milch cows, unaccustomed to hard labor—to become draft animals. What the men did was yoke their two cows and place them behind the lead pair of oxen and in front of the other pair. There the cows were expected to learn from the more experienced beasts, and apparently they did.

At the end of three weeks on the road Gurdon found time for reflection: "Whenever we struck camp it seemed like home whether on the Prairie, or in the Groves." Yet the journey was not always pleasurable or interesting: "And by this time the excitement began to wear off and scenes began to appear more every day alike." Here he seems to be articulating the feeling of the entire group, and at about this point in my reading I found myself thinking of the party as "G. P. Lester and Co.." I would later discover that Gurdon thought of it that way too.

The scenes that had come to seem so much alike were due to change very soon, for the company was approaching the Missouri. Here was where emigrants should outfit themselves for the mines, the guidebooks advised. Accordingly at Glenwood, shortly before they reached the river, the men bought picks and shovels. They had expected to meet Wakeman and Mann at this town, but when the two did not appear, the rest of the company drove on to the Missouri Bottoms, arriving about two o'clock in the afternoon.

At their back were high bluffs, evidence of an older, wider, deeper river; at their feet was thick, heavy grass, growing out of a fertile soil, which these farmers' sons could not fail to admire. They admired the trees, too, or at least Gurdon did. Most of them were cottonwoods, but with some sycamore, oak and honey locust interspersed. At the edge of the trees they camped and waited for Wakeman and Mann to rejoin the company. The next morning the pair drove in, and even though it was not Sunday they all took the day off, spending it "hunting and strolling around through the woods and along the river."

They got an early start the next morning and by nine o'clock were driving their wagons onto the steam ferry *Emma*. Resolutely, powered by her fiery boiler and her turning wheels, *Emma* ploughed her way through the swirling brown waters. Here was a great river, more formidable than any the company had seen since they left Clayton County. For these men, the Missouri formed the boundary between what they called "the states"—the safe, familiar, but somehow unsatisfying place—and what they knew as "the territories"—the land of danger and of promise.

After men, cattle and wagons were safely down the gangplank at Plattsmouth, the company went straight to Main Street, where the stores were. There they "made their stock of provisions complete," Gurdon says. He then saw an opportunity to buy some newspapers, and he let his companions drive on without him, thinking he could catch up to them quite easily on foot. Plattsmouth, with steamboats coming up the river from St. Louis, St. Joseph and smaller towns on the way, was a good place to look for newspapers. As a sample of the papers Gurdon might have picked up that morning, I looked at a copy of the *News* of Nebraska City for April 21, 1860. This town, twenty-five miles below Plattsmouth, was only ten miles upriver from the slave state of Missouri.

Of greatest interest to Gurdon, I imagine, would have been the Pike's Peak stories; there were six of them on page two, all suffused with optimism. One correspondent writes that in his district of the Rockies "there is room for at least 10,000 men to profitably engage in mining." Another says that the average miner can expect to earn between eight and twenty dollars in a six-hour day. Elsewhere on the page an editorial ridicules the efforts of the Territorial Legislature to abolish slavery in Nebraska. It argues that the three or four persons "ostensibly held as slaves in Nebraska City" are "in a better condition" than most white servants in the North. It gave me a shock to see this classic pro-slavery argument applied to my home state, and to a town I had visited as a child. Until then I had cherished the belief that Nebraska had escaped the stain of slavery.

The search for the papers so absorbed Gurdon's attention that he forgot to notice which road his companions took as they drove out of town. He walked for hours, eventually realizing he had taken the wrong road. At ten o'clock he "put up" for the night. "Early the next morning I was again in search for my company and at 7 oclock A.M. I again arrived at camp where there was a general laugh."

Though the company planned to link up with the Platte River Road, they did not begin to follow the river at its mouth. The Platte, as it flows through eastern Nebraska, makes a great northward bend before turning south and entering the Missouri. Knowing they could save many miles by cutting across this bend, the men traveled briefly south, then due west and finally northwest on the main branch of the Oxbow trail. On the night of the twenty-sixth, their second night in Nebraska Territory, they camped on the open prairie; the next day they crossed a creek named Salt; on the day after that they were hit by a blinding sandstorm. By the end of that day, glad that the sabbath had come, they pulled in to Pattents Ranch on Wahoo Creek, where they found thirty wagons stopped for the night.

Most of the other wagons went on their way that Sunday morning, their drivers evidently impelled more by their eagerness to reach their destination than by any concern for their souls or their cattle. The Lester company, however, remained in camp as was their custom. But there was no sermon that Sunday; indeed, Reverend Mann seems to have preached no more sermons for the rest of the trip. What the men did was wash their clothes, doing so, Gurdon says, "in rather of a rude manner," for the only washtubs they had were the pans they cooked in.

The most notable sight that day was a wagon train that passed them by. It was composed, like their own, of three wagons, but in place of oxen these wagons were being pulled by horses—twenty-six of them. The six men in this extravagantly equipped outfit were "enrout for California," Gurdon reports, with what sounds like a touch of envy. The caravan moved on, while the wind blew through the camp and "the prospect of rain was quite flattering." But the flattery was deceptive; the looked-for rain did not come.

The men were finding that timber was scarce on these prairies and grass very poor; they calculated that they had come sixty miles from the Missouri and still had thirty-five miles to go before they would reach the Platte. On May 2, an hour before noon, they arrived at Platte Bottom, the alluvial flatland that bordered the river. It had taken them exactly a month to come this far, and the stream itself was not yet in sight.

Three days later they camped for the first time at the edge of the Platte. No river they had seen in their travels had been quite like this one, its width measured in miles, its depth in inches, or so it was said. Actually it was deeper than folklore would have it, three or four feet at this place. The common assessment of its width was about right, though; a mile or two was the way travelers usually reckoned it. Its peculiar yellow color was borrowed from the sandy ground through which it flowed and the quantity of sand it carried in its waters. Its banks were low, so that the surface of the river lay only a few feet below the valley floor.

On Sunday May 6 they were joined by a friend of Russell and Wakeman, a man named Elder. "We also found A. N. Rich & Co.," Gurdon writes. That night the two companies, their members linked by ties of friendship and family relationship, camped together beside the broad and shallow Platte.

They spent another day together before moving westward again. About eight o'clock on Tuesday morning they began to see on the road ahead the tall flagpole that told them they were approaching Fort Kearny. Gurdon took advantage of the fort's mail service: "I stoped here about 1/2 hour. Mailed a letter to my friends." The entry provokes in me a strong curiosity. Did Seth mail a letter too? Did he send a message to Amorette, telling her where they had been and what they had done? Did he ask about the baby? Did he by any chance have a letter from her waiting at the Fort Kearny postoffice to be picked up when called for? On these questions Gurdon's diary is no help.

The May 22 entry recounts the second of several episodes in which individual members get lost and come to grief. Here Gurdon repeats a story Seth and George Wakeman have just told on themselves. The two of them "started at 4 oclock P. M. on a hunting excursion and got lost from the company and after wandering until about the hour of 12 (midnight) they were under the painful necessity of begging loging at the tents of some stranger. They were finaly received after asserting in bold terms that they were guilty of no crime, and pawned their guns to the prosprietors of the tent to guarantee good behavior. They came into camp about sunrise the next morning." Where, I suppose, they were met with a general laugh. I have to laugh myself, as my grandfather and his companion assert their rectitude "in bold terms" and then meekly consent to give up their guns in exchange for a space on the floor of a strange tent.

This story of a hunting excursion gone awry gives me no reason to think that Seth and Wakeman actually brought back any game. Small game would have been what they were looking for. Buffalo were out of the question, though the company was in the region marked on guidebook maps as "buffalo country." Seth and Wakeman had no horses with them, and even if they had, a buffalo could usually outrun a white man on a horse. Furthermore, no live buffalo were to be seen. "In the elaps of four days I think I saw more than 1,000 dead Buffalo," Gurdon writes. "But as yet have not seen a live one." He explains their absence by saying that the season was "quite backward," the grass was "very poor" and there was "nothing to induce them to come to the Platte Valley."

Gurdon's description of the dead buffalo may be misleading. Richard E. Jensen, who co-edited Gurdon's diary with me, believes that what Gurdon saw were not "carcasses of buffalo but skeletons left by hunters over many decades." Gurdon's other observations were accurate, however. 1860 was another drought year on the plains, as 1859 had been. For the emigrants, the drought caused inconvenience—grass was sparse and water scarce, the cattle suffered. But among the Indians, all the living creatures suffered, human beings and animals alike.

The company was soon to meet some of those hungry Indians: "We were visited today by two Sioux Indians, who came for the express purpos of begging. They had a letter of recommendation signed by different individuals stating that they were in very poor circumstances and greatly stood in need of something to eat. We gave them something to eat and then they shook hands and went away contented." The "different individuals" might have been traders, storekeepers, Indian agents or any other whites who were sympathetic to the Indians' needs. The handshake seems to indicate a certain degree of tolerance between the Iowans and the Sioux. But the contentment Gurdon thought he observed could not have gone very deep.

Shortly after the company got started again, Wakeman and Mann had a breakdown, and half a day was spent in putting a new axletree into their wagon. The delay was worrisome. More and more as the company pushed westward, water and grass were their greatest concerns. On Thursday Seth and Gurdon and Elder "waded the Platte in search of grass and game." They found a little of both, Gurdon says, "but not in paying quantitys." Already, though they are still many miles from the goldfields, he has picked up some of the miners' lingo.

By the end of that week the men concluded that they would have to travel on, forgoing their customary rest, in search of feed for their hungry animals, but four miles up the road the grass became more plentiful, and they stopped for what remained of the sabbath. They were now on the south fork of the Platte, which comes down from the place in the high Rockies to which they were heading. The going got harder from one day to the next; Gurdon's entries for May 21-23 speak of nothing but sand. On the twenty-third they had to double teams to get through it—apparently the first time they had had to resort to this tedious practice. Doubling teams meant that one team, the Foot-Lester team, for example, would be unhitched from its wagon and harnessed with the team of another, the

Bensons' perhaps, to draw the Benson wagon through a difficult section of road. Then the two teams would be unhitched and taken back through the sand to be harnessed to the Foot-Lester wagon and it would be pulled through. Finally one of the teams would be led back to the Wakeman-Mann wagon to help its team struggle through the sand.

The next day offered a respite—"good roads and very good grass"—but the day after that was the worst yet. When they approached the Cutoff, a newer road that left the South Platte and struck off diagonally toward Denver, they did not take it. Instead they stayed on the old road, which followed the river past the mouth of Bijou Creek. But they probably regretted their decision, for the three-mile stretch of road between the Cutoff and Bijou Creek was deep in sand most of the way. At four o'clock in the afternoon, when they reached the creek, there was nothing to do but allow the exhausted cattle to rest. At seven in the evening they started on again, determined to reach Fremont's Orchard, ten miles ahead, before they stopped for the night. But for half the distance, Gurdon says, they encountered "tremendous heavy sand," and it was one o'clock in the morning before they came to Fremont's Orchard, where they camped for the remainder of the night.

After such a day things would almost have to get better, and they did: "This morning we for the first time behold the Rocky Mts. We were told by those that traveled the road before that they are full 100 miles off whereas they don't appear to me to be more than 10 miles off. We can see clearly that these mountains are covered with snow. We can plainly see Long's Peak as it towers above all the adjoining mountains. The Black Hills can also be seen from this point." The wall of mountains, gleaming white in the morning sun, calls forth Gurdon's best powers of description, as he strives to make his language worthy of his subject. Reading Gurdon's words, I share his excitement, remembering the first time I beheld that same sight.

At Fremont's Orchard—not exactly an orchard but a grove of native cottonwood trees—the men were in what is now northeastern Colorado. The Black Hills that they saw, off to the north, were not the Black Hills of South Dakota. These Hills, called the Laramie Mountains on today's maps, were in what is now Wyoming; they took their name from the dark green pine and scrub cedar trees that covered them. Seth will see them again, day after day, for most of another summer. But today, Saturday, May 26, 1860, he turns with his companions toward the plains.

As the men make their way toward Denver, the watercourse that guides them, and which has been flowing east for most of their journey, is now flowing from south to north. They are in Kansas Territory; as yet there is no territory named Colorado. The mountain range, stretching across the western horizon, continues to fill Gurdon with admiration: "30th This morning the mountains look beautiful as the sun shone upon their sumits that was caped with snow." On the morning of the thirty-first they got their first sight of Pike's Peak "far to the southward." Someone told them that the Peak was 150 miles away, but that was an exaggeration. The real distance was somewhere between eighty and a hundred miles. The question was academic, of course, since they did not intend to go there.

At a little after noon on June 1 they reached Denver City, a town of "full 3,000 inhabitance," which Gurdon found to be "in a flourishing condition." But the high prices for all commodities (flour, for example, selling at three times the price it commanded in Iowa) sent them on their way before the day was over. They crossed the river and for the first time camped on the west side of the north-flowing South Platte. By three o'clock of the next day they were in the mountains, and at five they set up their camp in a "beautiful valley," where they found "very good grass" for their cattle and "the best of spring water."

The next day, which was Sunday, was uplifting to Gurdon's eye and to his spirit: "3rd This morning I took a stroll upon the mountains and the scenery is of a most beautiful character. I could look onto the Snowy Range and see both Long's Peak and Pike's Peak one far to the right and the other to the left." In the afternoon he had the opportunity to observe the sabbath in what for him was the proper way: "At 2 p.m. I listened to a sermon delivered by Rev. Mr. Rolls of the Methodist order. He spoke from Luke 11 chapt. and 28 verse. This was the first sermon I have heard since leaving the states. Our seats were ox yokes wagon tongues &c."

Luke 11:28 reads as follows: "Blessed are they that hear the word and keep it." Hearing the familiar biblical language in this unfamiliar land must have been reassuring to Gurdon. But was Seth sitting beside him, listening to the word, as his wife and his father would have wished him to do? Here again Gurdon's focus on himself—natural enough in a diarist—prevents us from learning the answer.

Chapter 5
Amorette's Baby

In the middle of June the census-taker came to Mary Rich's house in Westfield, Ohio. He—I assume it was a man—took down the information in a fluent, legible hand, but he was not very careful about particulars where the youngest member of the household was concerned. Listed beneath "Mary S. Rich, Age 59" and "Amouret Foot, Age 21," this member was registered as "Anglo, Age 2/12, Sex F, Place of Birth N.Y.." The age, two months, was the only thing he got right about Amorette's baby. The truth was that the little boy was born in Ohio and his name was Delizon Angello, or Delezon Angelo; the Rich Family Record spells it both ways. In place of either one of these names, I am going to call him Dell, a nickname I know he used when he grew older.

Out of the clothing that Amorette and her mother must have made ready for the baby's arrival—the blankets, the diapers, the dresses and wrappers, nightgowns and caps—one dress still exists. Amorette kept it with her always, despite the numerous moves she made in her adult life. I have it now. Though marred slightly by a few spots and small rips, it is still lovely. It is made of a very soft white cotton, so light that it seems to have almost no weight, and it has been painstakingly sewn and trimmed. The plain square neckline is drawn up in front and back by fine cords ; the sleeves are trimmed with lace. At a baby's knee level two wide tucks make bands around the long full skirt; below them come rows of narrower tucks. Along the lower edge tiny triangles form an engaging zigzag pattern. Someone, whether it was Amorette or her mother or some friend or relative, had an eye for beauty when she made this dress.

The most important concern for a new mother, then and now, has to be her child's feeding. I am assuming that Amorette breast-fed Dell. Most mothers in antebellum America did, or so it seems, looking back, though no one knows for sure. Scholars who have written on the subject indicate that although some mothers employed wet-nurses or resorted to nursing bottles, the majority nursed their babies. At that time, when substitutes for mothers' milk were primitive and refrigeration was unreliable, breast-feeding was the safest method by far, and the least expensive. In many ways it was the easiest, for some women it was pleasurable as well, and it had the authority of long tradition behind it. It seems a fairly safe assumption that Amorette nursed her baby.

When I had my first child, the assumption was that a mother would not nurse her baby. In March of 1946, at St. John's Hospital in Santa Monica, I was the only woman in my four-bed ward who had chosen to do so. In that first year after the war ended, with mothers and babies filling the hospital to capacity, our overworked nurses did little to encourage breast-feeding. The nurse who took care of me obviously felt it was a nuisance to bring my baby out to me at feeding times, when he could be fed more efficiently back in the nursery. Once I heard her say jokingly to him, as she carried him out the door of our ward, "*Now* you'll get your dinner." It's true that he needed special attention. At birth he had weighed only five pounds, three ounces, and in the next day or so, before my milk came in, his weight dropped below five pounds. But I was determined to be the one to nourish him.

When he got up to five pounds again my husband David and I brought him home. The doctor had instructed me to put the baby on a three-hour feeding schedule, instead of a four-hour one, because he was so small. Mother came to help me for a few days. She had taken time off from her hospital job, where she cared for new mothers.

The baby cried a great deal after he came home. Mother, noticing my breasts were full and dripping, decided he was crying because he was getting too much milk. She said that in addition to maintaining the three-hour schedule, I should limit him to ten-minute nursing periods. I followed her instructions. I watched the clock. My baby, knowing nothing about clocks, tended to wake up about two hours after his latest feeding. He would turn his head from side to side, opening his mouth, blindly searching for something to suck. Soon, if the nipple was not forthcoming, he would begin to cry. I would let him cry until the clock gave me permission to feed him, and when the clock showed that he had nursed ten minutes, I would put him away from my breast.

Mother had been unable to nurse me successfully when I was a baby. I had heard the story more than once: how she tried and tried but the milk wouldn't come. How Daddy, a doctor, watched me grow paler and more hollow-eyed day by day, until finally he said to my mother, "That child is starving!" How a wet-nurse was brought in but was soon deemed unsatisfactory. How various formulas were tried until at last one was found that I could digest—evaporated cow's milk, sterile water and Karo syrup. Knowing the story, knowing that Mother had had no more children, I knew that her advice to me could not have been grounded in personal experience. Still, she had the authority; she was a nurse; she was my mother.

David's aunt, a mother of five children, came over one day to see the new baby. He was, as usual, crying. Embarrassed and apologetic, I explained that he was crying because he was getting too much to eat.

She looked hard at the tiny, screaming boy. Then she turned to me. "That baby don't look like he's gettin' too much to eat," she bluntly told me.

Her words shocked me into seeing. I knew all at once that I had been looking at the whole thing backwards. The baby wasn't crying because he was getting too much to eat, he was crying because he was getting too little. Mother had been wrong. Small and underweight as he was, I should have been feeding him more often, not less, and spending more time, not less, at each feeding. I took Aunt Helen's advice and began supplemental feedings of formula, while continuing to nurse him as well. They were a nuisance and probably not necessary—probably I could have built up my milk supply again just by frequent nursings, but I did not know that then. In any case, I gave him all the milk he wanted, whenever he wanted it, and he stopped crying and began to grow.

Before my next baby was born I read Dr. Spock and learned what I should have done with the first one. He said that a new approach, called "self-demand," had recently been developed as a reaction against the rigid feeding schedules which had been recommended in the early years of the century. (Those were the years when Mother had received her training as a nurse.) Spock advised a middle course, beginning with great flexibility and then, for the convenience of the mother and other family members, gradually encouraging the baby toward a regular feeding schedule. But I liked the sound of the self-demand plan so well that I followed it wholeheartedly with my other babies: when they woke up and cried, I nursed them for as long as they wanted to nurse, and they thrived. No bottles were needed. It had been a hard lesson, however, and I have always wished that I had not learned it at the expense of my dearly loved first son.

Now, some sixty years later, as I consider Amorette's experience in light of my own, I find myself wondering how often she fed her baby. Where did she and her contemporaries stand on the question that so exercised doctors, mothers and on occasion grandmothers and aunts in the 1940s? I have found an answer in the writing of a nineteenth-century woman who, though not a mother herself, spoke with assurance on infant care, as well as most other aspects of what she called "domestic economy." This was Catharine Beecher, Harriet Beecher Stowe's older sister, whose *Treatise* on that subject, first published for national distribution in 1842 and reprinted numerous times in the ensuing years, made her name known across the country.

In her discussion of the feeding of infants, Beecher quotes two doctors whom she considers to be authorities. Their words might almost have been written by adherents of the self-demand school, one hundred years later: "'At that early age, there ought to be no fixed time for giving nourishment. . . The active call of an infant is a sign which needs never be mistaken.'" At the end of the quotation Beecher, speaking in her own voice now, warns against following this advice too slavishly: "After a child has satisfied its hunger, from two to four hours should intervene before another supply is given." Still, she treats babies less mechanistically than did the authorities who prevailed in the early years of the next century.

The second most important concern in a new mother's life is, or was, the diapers. I must add the disclaimer now that a diaper, a construction of cellulose fibers and plastic, is used once and thrown away. But for most of American domestic history, diapers were made of cloth and they were washed and put back on the baby, over and over again. Birdseye, a sturdy white cotton woven in a pattern of small diamond shapes, was traditionally used. I had some of the large, square birdseye diapers in 1946, though they were old-fashioned even then. But at least they were hemmed, ready to be put on the baby, when I bought them. Amorette and her contemporaries would probably have bought the material and cut and hemmed the diapers themselves.

Emily Hawley Gillespie, a young Iowa woman who was expecting her first child in 1863, noted in her diary the purchase of eight yards of "diaper." Clearly her baby's diapers were going to be made at home, probably by Emily. Even more interesting to me is her use of the word *diaper* generically. It tells me that a certain type of cloth (quite possibly birdseye) was so commonly used for diapers that it went by that name.

It is surprising, therefore, that Catharine Beecher's *Treatise* does not mention *diaper* or any word that might signify diaper, though it contains, in addition to the chapter on infant care, a detailed, prescriptive chapter on doing the washing. Perhaps Victorian prudishness caused Beecher to shun this vital subject, but I think, too, that she saw no need to give special instructions for diaper-washing because she saw no need to keep a baby's diapers separate from the family laundry, as my generation of mothers was instructed to do.

Early in her chapter "On Washing" Beecher enumerates the nineteen kinds of things that will be needed for doing the family wash—among them, four washtubs and "a brass or copper kettle, for boiling clothes." A "wash-stick" will be required for stirring the clothes while they are being boiled. A washboard, naturally, will be indispensable. The steps in Beecher's washing process amount to nine, beginning with an overnight soaking of the clothes. The next morning the washerwoman, or the lady of the house if she is the washerwoman, sorts them according to color and fabric and washes them in first and second suds, using two of the washtubs and of course the washboard. Then she boils them for half an hour. The next step is optional: she washes, for a third time, those articles that are still stained. Then she rinses all the clothes, puts the white things in blueing-water and starches articles that require stiffening. Finally she hangs her washing on the line to dry.

I envision Amorette performing some, but not all, of these tasks. One or two of the washtubs can be dispensed with, and certainly she will not want to blue and starch her baby's diapers. I will probably have to allow her to put the diapers in with the rest of the wash, though I can't approve of the practice, but then I will have her boil the clothes as recommended, which will remove my objection.

My mental picture of the young Amorette, bent over the washboard, scrubbing the stains from her baby's diapers, calls to my mind a picture of my younger self doing the same thing. I am on the porch of our apartment, beside the kitchen door, and I am standing before a large stationary laundry tray, which is partitioned down the center to make two washtubs. I have soaked the diapers overnight; I have boiled them because Dr.

Spock said boiling would kill the bacteria that cause diaper rash; I have drained off the water into the lefthand tub and filled it with clean hot suds. Now I am scrubbing each diaper up and down the ridged steel surface of the washboard. The reason I am doing the washing this way is that our landlord, and more particularly his wife, do not want to look out the back windows of their house and see a washing machine on the porch of our apartment. Since David and I have already broken Mr. Phillips's no-children rule, and since it would be next to impossible to find another apartment, given the postwar housing shortage, we think it unwise to defy him a second time.

After rinsing the diapers several times I would hang them to dry on the pulley clothesline that ran from a wooden pillar on our porch to a beam on the Phillipses' house. I did this six days a week, Monday through Saturday. Mrs. Phillips insisted I was not to hang out laundry on Sunday. I obeyed, resentfully; I didn't like to let the diapers accumulate for two days. Here, then, is one difference between my recollected picture of myself and my imagined picture of my grandmother. I don't think that Amorette, a staunch Methodist, would have considered doing the washing on the Lord's day.

Of course the greatest difference between Amorette's experience as a new mother and mine was that my baby's father was present to love and support us both. During my pregnancy David had given his unborn child a pet name, Bumby, and the name stayed with the baby for the early months of his life. Piercing the clouds that drift across memory, forgetful of the shocks, the rupture and the grief that were to come, I see again the bedroom where the three of us slept. I see the two dressers, bought unfinished, sanded and stained by me; the double bed with the slatted headboard that David had painted blue; and the fine new crib that filled the space before a sunny window.

At first Bumby seemed almost lost in his crib, tiny as he was, and he often finished the night in bed with us. I think he felt comfortable nestled between his parents.

I feel sure that the infant Dell never lay beside his father, never felt the warmth of his father's body, never looked into his father's face. In June of 1860, while Amorette and Mary were talking with the census-taker in Ohio, Seth was in Kansas Territory, twelve hundred miles away. Fourteen months later, when he joined the army, he gave a town in Kansas as his place of residence. Reluctantly I have concluded that Seth remained in the West during Dell's infancy.

This conclusion is hard for me to accept. Surely, I want to say, he would have gone back to see his son. But the evidence I have of this part of Seth's life indicates that he did not. The consequence for Dell was that he knew only one parent, his mother.

Necessity laid a solemn obligation upon Amorette, for nineteenth-century America regarded motherhood as a high calling. Catharine Beecher had dedicated her *Treatise* to American mothers, paying special tribute to the mothers of sons. The mother, Beecher wrote, "forms the character of the future man." In the spring of 1860, in her mother's house, Amorette entered into her life's work.

Chapter 6

Testing Their Claim

On June 4 the company got under way again, all of them having benefited from the day of rest, but they soon learned that mountains, no matter how beautiful, can cause difficulties for the traveler. "This morning we commenced climbing the mountains," Gurdon writes, "and longer and steeper hills I never had the exquisite pleasure of ascending." The steepness of the climb and the roughness of the road were proving too much for the already weakened underpinnings of Wakeman and Mann's wagon. They broke another axletree on Tuesday, and another half day was lost in replacing it. That evening the company drove their wagons across the river and camped on the east side.

In the morning Gurdon tried his hand at hunting and made a comic story of his failure. "I had an excelent shot at a wild turkey," he says, "but owing to my being deseased with the buck fever I lost him, quite fortunate for the turkey." The comedy came to an end, however, when the men got out on the road. "Today the roads are still worse than ever as they are very stony and full of stumps." The wagons had to cross the river six times that day, each time lurching down the bank, jolting over the stony riverbed and up the other side. Predictably, the newly repaired wagon could not withstand the stress: "Wakeman & Mann again broke down and as there was no feed we came on and left them behind." That is the last reference to Wakeman and Mann in Gurdon's diary.

For the rest of the company, matters improved on the next day. The roads were "a considerable better," Gurdon writes. By noon they had arrived at South Park, an expansive, stream-fed basin with a line of low mountains rising at its far edge, and behind them, the snow-covered peaks and ridges of a more distant range. Knowing they were close to their journey's end, they pushed on another five miles before stopping for the night. The next morning the company, diminished now by two, got an early start. At five p.m. they arrived at Fair Play diggings, their destination. That evening Gurdon "washed several pans of dirt," in which he "got the cullor of gold very readily." But there was a difference, as the men would soon learn, between getting the color and getting gold "in paying quantitys."

The mining camp of Fair Play lay along the South Platte. Up here, at an elevation just under 10,000 feet, it did not seem the same river as the one the travelers had known down on the plain. It was much narrower, its banks were higher, its water clearer (even after the mining operations had stirred it up), its bed strewn with stones. A group of miners who were already there had dug a ditch, diverting the river and laying the stream-bed bare. Eager to become full-fledged miners themselves, the seven men in the Lester company bought a mining claim the next day, promising to pay $250 for it out of the earnings it was expected to bring them.

That same evening the diversion dam gave way, causing "a great damage to the ditch company," Gurdon reports. The collapse was bound to damage the prospects of our men's small company as well; still they continued on the course they had set themselves. The next day being Sunday, they kept the sabbath, and Gurdon heard another of Mr. Rolls's sermons. The snow that fell intermittently all day seemed to bring Snowy Range, ten miles to the northwest, closer than ever to where they were. On Monday morning, June 11, they started building a cabin, planning to live in it for the summer season. On Wednesday they finished it and moved in. Then it was time to get sluice boxes and test their their claim. Ten men—probably Seth, Gurdon, Henry, the two Bensons, Russell and Elder, along with three others whom they met in the camp—ran a sluice all one day. But when they took their gold to be weighed, they had earned a total of $3.87.

This, then, was what it had all come to. Two months of travel, the loss of two of their members, the building of a cabin "for a season"— after all that, each man earned thirty-eight cents for a day's work. They tried several other places, and were always able to raise the color but never in a paying quantity. Disillusionment set in with a rush: "Finally all became dissatisfied and proposed going to California some of our party are now making arrangements to start for California." The fact that Gurdon does not tell his diary which members of the party are making those arrangements suggests that he has not yet made up his mind what he will do. But his sense of grievance is clear. The beautiful mountain region, so thrilling to look at, has now manifested its ruthless aspect: "There has not been but one night since I arrive at Fair Play diggings but what it has froze water. It is said by those that pretend to know that Fair Play diggings are 8,000 feet above Denver City. We being at such an altitude the air is very light and let a person attempt to ascend much of an elevation he finds his breath nearly exhausted by the time he arrives at the top."

The persons who pretended to know the elevation of Fair Play exaggerated its elevation by nearly four thousand feet, but the rest of Gurdon's complaint rings true: "At the time I now write it is the 24th of June and there is scarcely any body that can work their claim to any advantage on account of high water. They can sink a shaft not to exceed 5 feet before they are prevented from going any further by water's coming in notwithstanding the river is turned and the old channel is well drained." Nothing about this place can be trusted, it seems. The people who claim knowledge about it don't know what they are talking about. Nature, paying no attention to the calendar, refuses to admit that summer has come. The watercourse you thought you had tamed asserts itself again. Fair Play does not play fair with those who come to share its bounty.

The twenty-fourth of June was a Sunday, and the Reverend Mr. Rolls preached to the miners again. It was the last sermon Gurdon would hear at Fair Play diggings. Two days later he and Henry Foot, along with four other men, set off for California. But Seth did not go with them, nor did Russell or Elder or the two Bensons. Certainly Wakeman and Mann did not go; as far as we can tell, they never even reached Fair Play.

Gurdon continued his diary as he went west, and the repeated concern he expresses over the poor quality of the grass makes me think that he and Henry took the team and wagon which the three of them had brought from Iowa. If that is so, where did it leave Seth? For a time I assumed that he and the remaining members of the company were as disillusioned with the Rocky Mountain gold mines as Gurdon was and that when Gurdon and Henry set out for California, the rest of them headed for home, becoming "go-backs." But then I discovered two pieces of evidence which contradicted that assumption. The first was the original mining claim, which I found at the library of the Colorado Historical Society. Here is where it becomes clear that Gurdon thought of himself as the company's leader:

June the 11th 1860
G P Lester an co Claim 100 ft in
Paid this District Down the Stream
Below the Line as No 41 H Henson
G P Lester Claims 50 feet in this District
Paid Down the Stream below the Line
the up half of 42 H Henson

I do not understand what "Below the Line" means or why Gurdon claimed that extra fifty feet, which he did not mention in his diary. But for my story the most important feature of the claim is the word *Paid,* written twice in the margin and obviously added later than the original transaction. What the additions mean is that after the self-appointed leader abandoned his company, the remaining members, or some of them, stayed on in Fair Play for a while, worked the claim, made some money and paid off the $250.

The second piece of evidence, which I also saw at the Colorado Historical Society, was the 1860 census for Kansas Territory. In it, three members of the company appeared. In fact Franklin Russell and the Bensons turned up more than once, in various locations and permutations. William Benson and E. Benson are listed at South Park and also at South Clear Creek. An F. Russell and an F. E. Russell are listed at South Clear Creek, a Frank Russell at Nevada Gulch. But Seth is not listed anywhere.

I believe that he, the Bensons and the Russells were still in Fair Play on June 26, the day Gurdon and Henry left. After that date I lose track of Seth for fourteen months. He comes back into view on August 28, 1861, at Fort Leavenworth, Kansas.

PART THREE

Seth's Story

Seth at Fort Leavenworth, 1861

Chapter 7

His War in the West

The day Seth was mustered into Company A of the Eighth Kansas Volunteer Infantry Regiment was the first day of the company's existence. Captain James L. Abernathy, who had organized it, was its commander. At that time Kansas was the newest state in the Union, having been admitted at the beginning of the year. Fort Leavenworth was on its easternmost edge, overlooking the Missouri River; on the opposite bank lay the slave state of Missouri.

The question of why this Iowa man enlisted in a Kansas regiment is not difficult to answer. He enlisted there because he was living there; his army records call Winchester, Kansas, his place of residence. But as his granddaughter I am moved to ask a larger question: why did he enlist in the army at all? Why, when he had a wife and child who were presumably his first responsibility, would he voluntarily commit himself to a service in which he might be killed? He was the only one of Moses Foot's sons to do so. Alfred, child of Moses' second marriage, was too young to serve. David, eighteen years old at the time Seth enlisted, never joined the army. He had married at seventeen and by war's end was the father of two children. Henry and his stepbrother Gurdon stayed in California until the war ended.

Yet Seth joined up, for what he must have considered at the time to be good reasons. Probably he felt, as many men did that August, that it was his patriotic duty to serve his country and help save the Union. President Lincoln had just issued a new call for volunteers. His first call, issued in April when the war began, asked men to serve for only three months, since it was at first thought that the rebellion would end in no time at all. The disastrous Battle of Bull Run, in July, put an end to that way of thinking. In August the three-month enlistments were expiring and new recruits like Seth were required to commit themselves to a three-year period of service.

He might have thought that service with Company A would be comparatively safe, since the Eighth Kansas was being called a home guard regiment, formed for the express purpose of defending Kansas and the regions adjacent to it. The twin threats of Indian attacks along the Nebraska border and the Confederate presence just across the river were worrisome but did not approach the level of danger that faced soldiers in the theatres of war to the east.

Another possible reason for his enlistment is that the army offered him a steady job for three years. Knowing what I do of his financial difficulties, I am guessing that the promise of a monthly wage had some influence on his decision. While his army pay would be low at first, he could look forward to promotions, and in any event his living costs would be paid. Perhaps he might be able to send a regular allowance to Amorette.

She was living in Iowa again. She had come back, bringing Dell of course, and Mary Rich had come with them, leaving her home in Ohio permanently. But Amorette did not return to Clayton County this time; instead the three of them traveled on to Fayette County, where Angello was now living. Quite possibly Angello had encouraged his mother and sister to join him in "the beautiful land." The two women and the little child would make their home in the town of Fayette. That is where Seth would send his letters for the duration of his army service. . . . How naturally the phrase emerges from a vocabulary I had put put away over sixty years ago. That is what we said then: people made makeshift, temporary arrangements "for the duration"—doubling up with relatives, taking uncongenial jobs, just getting by until the war would be over.

At the end of October Seth was promoted to the rank of first sergeant. Two months later, when Second Lieutenant John Conover was chosen to organize a new company, the vacancy thus created gave Seth the opportunity for another promotion. On December 12 he became Second Lieutenant Foot. Not long afterward he went to a photographer's studio in Leavenworth City and had his picture taken. In the photograph he is wearing a dark coat with a narrow standup collar. The buttons that run down its edge shine out against the darkness of the cloth. He wears his coat unbuttoned, showing his white shirt. This is the photograph from which the crayon portrait of Seth, the one that used to hang in our front hall, was made. I like the photograph better. It shows Seth to be a younger man, less remote, more human. In the steady gaze of his eyes I see a trace of boldness, a hint of challenge—or is it simply pride in the new rank and the new uniform?

That fall two men who would play significant roles in later

chapters of Seth's story joined the Eighth Kansas. In late October John A. Martin was mustered in as Lieutenant-Colonel; in November Sol R. Washer was mustered in as Sergeant-Major. Martin would soon be given command of the regiment; one of his responsibilities would be to report any casualties suffered by the Eighth. Washer, as its adjutant, would become personally acquainted with Seth.

In February 1862 something happened that must have been a disappointment to a good many citizens of Kansas and perhaps to some soldiers as well. The Army ordered Company A, which by that time was stationed at Lawrence, Kansas, to march to Fort Kearny, Nebraska Territory. There they would join a force of volunteers from Iowa and Nebraska. The volunteers were replacing the garrison from the regular army, who had been sent east to fight the Confederates. The transfer marked the end of the Eighth Kansas' identity as a home guard regiment.

Seth and his companions started on their journey to Fort Kearny on February 15. Probably they took the Independence Road, meaning that they would have had three rivers to cross—the Kansas, the Blue and the Big Blue—very likely frozen over in this hard winter, the coldest, it was said, in ten years. Once past the rivers, they would continue northwest, entering Nebraska Territory and ultimately joining the Platte River road that Seth and the other goldseekers had traveled nearly two years earlier. When they saw the tall flagpole that seemed to grow steadily higher as they approached, they would know that their long cold march was almost over.

Fort Kearney, when they reached it, "was not an inviting place. It was lying on a large low flat, much lower than the already low and level prairie over which we had marched for many days." The description comes from the memoir of another soldier, Sergeant James Larson, whose unit, the Fourth U. S. Cavalry, was already at Fort Kearny when the Kansas company arrived. The word *march,* in this context, does not necessarily mean that all the soldiers were walking. Larson certainly rode to the fort and Seth probably did too. Like the other officers, he was entitled to a horse.

Once they were settled in at Fort Kearny, the men of the Eighth could not complain of overwork. Though they had been sent there to defend the mail coaches and the emigrant parties from the Indians, the emigrants were not traveling at this season of the year, and no hostile Indians appeared. Yet Larson probably spoke for many other soldiers when he called Fort Kearny "the most lonesome and uninteresting spot on the face of the earth." The sounds that broke the quiet of the windswept plain were not the kind to cheer a man up. There was the wind itself, howling around the corners of the buildings, rattling the flagpole and setting the flag to flapping angrily. And every night and every morning there was the boom of the guns, reminding the men of the distant conflict that had, indirectly, brought them to this place. Seth and his comrades must have wondered at times why they were at this bleak outpost, with so little to do, but in late April that question became irrelevant for him. It was then that he was detached from his regiment and attached to Company H of the Fourth U. S. Cavalry, which was ordered to Fort Laramie in what is now southeastern Wyoming. Company H was Sergeant Larson's company.

Sergeant Larson was only twenty-one the spring he rode out to Fort Laramie. Some fifty years later, as he was writing down his memories of his wartime service, he described the journey. "The weather being very pleasant, no tents were pitched. On coming in camp we simply unsaddled our horses, watered and staked them out, and later on gave them corn and groomed them a little. When the time came to go to sleep, each man took his saddle for a pillow and spread his blanket on the ground and slept just as we did later in active service." To be released from the tedium of Fort Kearny and sent on an adventure like this would seem to have been a piece of rare good fortune for Seth. Possibly he came by it because he was known to have traveled this road, or part of it, before. The cavalrymen would follow the Platte River road to the point where the river's two forks join to make one stream. Here their route, diverging from the road to the Rocky Mountain gold fields, would turn northward. They would ford the South Platte to reach the North Platte and would then follow its course almost all the way to Fort Laramie.

Little by little, as the cavalry train rode across the plains, they were gaining elevation, and at length they began to see the dark outlines of the Black Hills. Seth would have remembered them from his previous trip, but this time they were closer. And another mountain came into view, one that he had not seen before, higher and brighter than the cedar-clad Hills. This was Laramie Peak, its snowfield gleaming white in the spring sunshine. The party continued northwest, led on by the shining Peak, until they reached the spot where the clear mountain waters of the Laramie River flowed into the sandy yellow Platte. Fort Laramie, built on a rise of ground beside the Laramie River, had a more imposing presence than their last fort; that in itself was good for men's spirits. It contained many more buildings, one of which was allocated to "Quartermaster and Commissary," and this must have been the place where Seth worked. His job at Fort Laramie was Acting Assistant Quarter Master.

On the face of it, it might seem strange that reinforcements were thought to be needed at Fort Laramie in the spring of 1862, when the danger that the western territories would fall to the Confederacy was largely over. But now there was another danger: hostile Indians, some of whom were attacking emigrant parties, the Overland mail routes and the newly completed transcontinental telegraph line. In the region under the Fort's jurisdiction, bands of Indians had destroyed mail stations and had killed or wounded stagecoach drivers, passengers and mail company employees. It was obvious to the Army that a stronger presence at Fort Laramie was needed.

I do not think that my grandfather fought any Indians while he was in the West. If there had been any fighting between the Fourth Cavalry and the Indians, Sergeant Larson would have recorded it; instead he recorded his disappointment at its absence. When his unit was first ordered to Fort Laramie, he said, they had looked forward to a "chance to have a lively spat with the warlike Sioux." But after they arrived, they "found that the Indian trouble had been greatly exaggerated."

Yet the danger was not imaginary. The commander of the fort, Brigadier-General James Craig, was in a constant state of anxiety about it. On June 26, he wrote his superior, Brigadier-General James G. Blunt, at Fort Leavenworth: "Indians committing depredations almost daily between Green River and Salt Lake. . . . What shall I do?" Six days later he wrote General Blunt again: "One hundred and sixty head of animals taken from emigrants at Ice Springs night before last. . . . I cannot cover 500 miles of road with the small force now here. . . . I am now of opinion that the Snakes, Crows, Cheyennes, and one band of Sioux have taken part in the depredations."

"Depredations"—The Official Records of the War of the Rebellion, where these letters of General Craig's are to be found, speak often of them, and always they have been inflicted by the

Indians on the whites. The records do not speak of the depredations visited upon the native tribes by the whites—the buffalo driven off, the streams befouled, the diseases introduced, the starvation, the death.

The prevailing anxiety about Indians led to a new assignment for some members of the Fourth Cavalry, and I think Seth was among them. Because the threat to the present mail route was so great, a more southerly route, thought to be less vulnerable to attack, was to be laid out. To protect it, a new fort, to be called Fort Halleck, would be built at the northern base of Elk Mountain, in the Medicine Bow range. In the middle of July an Ohio cavalry regiment was dispatched to Elk Mountain to build it. There they were met by a detachment of the Fourth U. S. Cavalry, who had ridden to the site before them. On July 14 Seth had been given a new assignment, that of "Assistant Commissary of Subsistence." It sounds as though he may have been placed in charge, or partially in charge, of the food supplies needed for the cavalry's expedition to Elk Mountain.

The upland meadow that was to be the site of Fort Halleck lay at the foot of two mountains, one behind the other. The lower one was a forested ridge, rounded off at its highest point, and sloping down gradually on either side. Behind it stood Elk Mountain itself. Entirely white, it outlined the contour of the ridge except at the summit, where it reared a sharper peak abruptly skyward. The men who came into this place must have felt dwarfed by it, overshadowed by the mountain and hugely outnumbered by the wild animals that roamed the land on every side. One of the Ohio cavalrymen described the wild creatures this way: "At the time we landed at Fort Halleck until we left, the vicinity was a vast grazing ground for thousands upon thousands of elk, antelope and black tail deer, in sight any time of the day, and the chaparral along the base of elk mountain was full of cinnamon bears. Scarcely a day passed without a flock of antelope chasing through our camp, upsetting tents, camp kettles, etc.."

Two years before, as Seth made his way into the West for the first time, he often went out hunting for game but rarely, if ever, succeeded in killing any. Now, it seems, the game was all around him. Did the Assistant Commissary of Subsistence ever take the opportunity of procuring some venison for himself and his companions? I see no reason why not.

It was probably August before Seth and the cavalrymen got back to Fort Laramie. August on the high plains shows signs that the summer is coming to a close—the nights are cooler and some of the leaves on the cottonwood trees have turned yellow. Sometimes a cold rain falls, as it did that year. In August of 1862 the Mormon wagons continued to pass through Fort Laramie, their drivers confident, apparently, that they could get over South Pass and down into Utah before winter set in. But emigrants whose destinations were California or Oregon should, if they were prudent, have passed through earlier. To the men just returned from the site of the new fort, the scene at Fort Laramie must have seemed quieter than it was when they left. It would have been a good time to write letters home.

If I could find the lost Civil War letters, I think I would see that Seth had many things to tell Amorette about his service at Fort Laramie: stories about the scenery and the wild animals, perhaps, and surely episodes that took place in and around the fort. But there was one episode I think he would have kept to himself. Knowing Amorette as I have come to do, I don't think he would have wanted her to know he had been gambling.

Sergeant Larson was the one who eventually committed that story to paper, quite naturally casting himself as its hero. Not long after the cavalry returned from Fort Halleck, Larson was given a special assignment, the capture of a horse-thief and the recovery of the horse: "It appeared that a gentleman gambler had come over from Denver City and by his good manners and smooth language, as well as his fine violin music, had been entertaining the officers. But in their social evening games at draw poker he had managed to empty their pockets also, and one fine morning he was missing, with a fine horse belonging to Lieutenant Foot."

By the time Larson received his assignment, the gentleman gambler had a head start of about ten hours. Wisely deciding that the man was on his way back to Denver, Larson guided his horse to an Indian trail which led in that direction. All that day he followed, and when night fell he bedded down beside the trail: "Daylight was just beginning to come the next morning, a slight tint of red began to be visible in the east as I was coming down a very long slope from where I could look into a broad valley where a few scattering trees were growing. It was then so light that I could discern objects at a distance, and soon I spied two horses staked out and not far from there under a small tree was a spread where apparently somebody was lying asleep."

To deaden the sound of his horse's shoes, Larson left the trail and made the final approach over the grass at its edge. Still mounted, he stood over the sleeping gambler and, with revolver drawn, called upon him to surrender. Having no other choice, the gambler did just that, and the two men began the return trip, slowed somewhat by the necessity of leading Seth's horse. As they rode along, the prisoner said he knew that trail so well that he "had considered himself safe, and in a position where no pursuit was to be feared."

And there, unfortunately, Larson's story ends. The sergeant says nothing at all about the restoration of the horse to Seth, and I am left to imagine my grandfather's feeling of relief, after two days of worry.

One man at Fort Laramie who had no respite from his worries, in the last days of August 1862, was General Craig. On the thirtieth he sent an urgent plea for reinforcements to General Blunt. Craig says he has heard from a reliable source "that several thousand Indians from the Upper Missouri are now approaching the post with the avowed intention of making war." He is so desperately short of men, he says, that he will accept a company of soldiers without either horses or arms—he can manage to arm them and find horses for them. Craig's superiors responded to his appeal by sending the Fourth Cavalry east to St. Louis. The needs of the real war had superseded those of a war that was only rumored and feared.

Seth left Fort Laramie with the cavalrymen. They rode east across Nebraska Territory to Omaha City and there went down the Missouri to St. Louis. He stayed with them for a month or so, returning to Company A at Fort Leavenworth in mid-December. But it was not long before the company would be ordered east. On February 22, 1863, they arrived in Nashville, Tennessee, where they would join the rest of their regiment in the occupation of that defeated city. There they would find that Colonel Martin, commander of the Eighth Kansas, had been appointed Provost Marshal, with the responsibility of maintaining order among civilians, military personnel and prisoners. John Conover, whose promotion in the fall of 1861 had allowed Seth to rise to the rank of second lieutenant, was now Captain Conover, Chief of Police.

On the army muster rolls for the months of January and February Seth is marked "Present" with no recorded absences.

But somehow, at the beginning of February, he managed to get back to Clayton County to transact some very important business.

Chapter 8

Seth Makes His Will

No matter what the muster roll says, I have good reason to believe that Seth was in McGregor, Iowa, on the second day of February, making his last will and testament. Quite possibly it was also his first; the war must have caused many soldiers to think for the first time of their mortality. Two witnesses were present, County Clerk Thomas P. Updegraff and a man named George C. Dibble. Both of them identified themselves as residents of McGregor. No other person is named in the one-page document that was being drawn up, but very likely a lawyer and his clerk were also present. Someone, probably the clerk, wrote out the text of the will on a sheet of paper, blue in color. He left space for Seth to write his name and below that space added a statement for the witnesses to sign. Seth wrote his name under the three provisions he had authorized, marking the import of the act he was performing by embellishing his capital S with a more intricate whorl than I have seen on his signatures elsewhere. Updegraff and Dibble signed their names in a business-like manner, *Dibble* neatly aligned under *Updegraff, McGregor* under *McGregor.*

Neverthess, the will was an inky document when all the men were finished with it. In the third provision the clerk had absent-mindedly written "I give" for the third time in succession. Then, realizing that he had already written all the bequests there were, he wrote *hereby* on top of *give* with such a heavy stroke of the pen that the ink smudged and ran over that word and the next one. There is a dark thumbprint in the upper lefthand corner of the paper and another one halfway down the left margin. These marks, however, do not prevent the completed statement from clearly expressing Seth's intentions.

His will is only four sentences long. The opening sentence indicates that in spite of his sojourn in Kansas and his membership in a Kansas regiment he continues to think of Iowa as his home: "Know all men by these presents that I Seth Foot of Clayton County Iowa being of sound mind and memory do duly make and declare this to be my last will and Testament." Next come the three provisions.

His bequest to Amorette follows the principle of dower rights, as it had been incorporated into Iowa law: "1st I give devise and bequeath unto my beloved wife Amoret E. Foot one Third of all my Estate Real Personal or mixed which is intended to be lieu of dower." Most often a widow's dower right applied to her husband's real estate, of which she would be entitled to one-third during her lifetime. In addition, she would be given, permanently, one third of his personal property. Since Seth did not own any land, Amorette's third of his estate would be limited to articles of personal property, such as clothing, household goods, keepsakes and of course money.

The provision for Dell sounds more generous than the one

for his mother: "2d I give devise and bequeath unto my beloved son Delazon Foot the remainder of all my Estate wether Real personal or mixed." But Seth was not showing favoritism here; he was merely following Iowa law, which required a husband to bequeath the remainder of his estate to his child or children, after the widow's one-third had been allotted. Young as he was, Dell would need a guardian to manage his property during his childhood and youth. The guardianship, though not mentioned in the will, would devolve upon Amorette, for she was given that right by law "if the court deemed her a suitable person," as she undoubtedly was.

Thirdly, Seth gave his father the responsibility of transacting all the legal and financial business that the carrying out of these provisions would entail, appointing him to be the "sole Executor of this my last will and testament." Then came the date and Seth's ornate signature and finally the avowal of the witnesses "that the foregoing Will of Seth Foot was executed in our presence by said Seth Foot and we at the request of the testator and in his presence and in the presence of each other have signed our names," leaving no doubt that Seth was there in the room with them.

I visualize this scene as taking place in one of the narrow red-brick buildings that still stand, carefully preserved, on McGregor's main street. The room, as I envision it, is fairly crowded. Beside the five men already mentioned, Moses Foot must also have been present, for another matter of business was transacted that day: Moses borrowed from Seth the sum of eight hundred dollars, giving his son a note for that amount. The existence of this note raises several questions. Why did Seth loan his father eight hundred dollars? Was it to assist Moses in some urgent financial need, perhaps to satisfy a debt on his land? Or was the transaction not a real loan at all but only a way to shelter the money from the claims of Philander Austin?

Seth must have known that he still owed his former partner a considerable sum, though Asaph Tuttle had made strenuous efforts on Austin's behalf in the spring of 1860—taking over a mortgage from Seth, repossessing his merchandise and pursuing him over the country roads of eastern Iowa. In the nearly three years since then, Seth had accumulated this eight hundred dollars somehow, but he had shown no intention of paying off the balance he owed Austin. Now that Seth was about to enter what he knew was the theatre of the real war, did he make this "loan" to his father in the hope of keeping the money out of Austin's hands and thus saving it for Amorette and Dell? Did Moses agree to cooperate with the plan? And where did Seth get eight hundred dollars? His army pay would never have allowed him to save that much. Perhaps he did find gold in paying quantities.

The undoubted fact of Seth's presence in Clayton County that February day does not explain his absence from his company. Not only is he marked present on the muster roll, there is no furlough order for this time period in his military file, though there is one for the furlough he received a year later. It is possible that he found a way to make a short trip home without official leave, but such an infraction does not seem consistent with the rest of his army record. It is also possible—and I incline to this explanation—that he did obtain permission but that the record was lost. The discrepancies and omissions I put down to clerical error, of which I have found other examples.

There still remains the question of the weather. The steamboat was the logical way for a soldier to travel from a fort on the lower Missouri to a town on the upper Mississippi, but were these rivers navigable at that time of year? They may well have been. *The North County Weekly Times* (published in McGregor) reported on January 21 that a warm spell had set in shortly before that date. Up until then the ice skating on the river had been "splendid," but since then the warmer temperatures had made skating impossible. The steamboats may have taken advantage of the melting ice to begin their runs again.

Obviously my answers to the last two questions are conjectures; somehow Seth got home. But to the question that touches me most deeply—was Amorette present at the signing of the will?—I can give no answer. Common sense suggests that she was not. To travel from Fayette to McGregor on the muddy roads of a winter thaw would have been difficult, perhaps impossible. Furthermore, the making of a will was thought to be men's business. But in spite of these obstacles I would like to believe that she was there in in the room with Seth. He called her and Dell by the most endearing of terms: *my beloved wife, my beloved son.* These were the conventional phrases, for testators who were husbands and fathers, but I would like to think that after the years of separation Amorette heard Seth—in his own voice, in her presence—say them aloud.

Chapter 9

The Colonel's Aide-de-Camp

On the Fourth of July, 1863, Colonel Hans Heg was in Brigade Headquarters at Winchester, Tennessee, writing a letter to his wife Gunild. It was the day General Lee lost the battle of Gettysburg and Vicksburg surrendered to General Grant, but Colonel Heg did not yet know of either of these momentous events. The most important item in his letter was the dinner party General McCook had put on that day. Heg himself had been one of the guests, and General Rosecrans had sat at the head of the table. . . . It had rained that day as usual, and Heg's men were taking the opportunity to rest after a muddy, uncomfortable march from Murfreesboro. . . . He was sending Gunild a picture to add to her collection. The picture, he wrote, "is that of a young fellow, who acts as my Aid De Camp." Already the new aide had become a great friend of Albert Skofstad, a staff officer in whom the colonel placed much confidence.

The aide-de-camp was Seth, and the picture must have been a copy of the photograph he had had taken at Leavenworth. He was one of some five hundred aides-de-camp who served in the Volunteer Army of the United States during the Civil War. They assisted their superior officers in a wide variety of ways, essentially doing whatever the superior needed to have done. This new appointment qualified Seth for a stipend of twenty dollars a month in addition to his second lieutenant's pay.

Working at Colonel Heg's side, doing his bidding, running his errands, Seth came to know his commanding officer quite well. Heg was a Norwegian immigrant who, before being given command of the brigade, had commanded the Fifteenth Wisconsin regiment, composed almost entirely of Scandinavians. Though he had called Seth a "young fellow," he was only four years older than Seth. He looked considerably older, however, with his full beard and his soldierly bearing. Outwardly he maintained a reserve that sometimes bordered on sternness, yet inwardly, as Seth must soon have realized, he was a man of strong feelings: love for his wife and children, love for his adopted country, ambition for himself.

The amiable, generous side of Colonel Heg showed itself most plainly in the evenings, when one or other of the two brass bands in the brigade would come by his tent to serenade him. After the concert he would reward the musicians with a glass of beer. He and one of his regiments got into a minor skirmish with the enemy one day but nobody got hurt. And so the summer at Winchester went by, with a good deal of entertainment and very little danger. By the end of August Seth had been in the army two full years without ever having participated in a battle.

The second anniversary of his enlistment found him and Colonel Heg and the rest of the Third Brigade on the west bank of the Tennessee River, preparing for the crossing. They were at a place called Caperton's Ferry, though for them there was no ferry, nor any bridge, either. The engineers would have to build one, putting down a row of pontoon boats, fastening one to the other as they worked their way across the wide, swift stream. Then they would lay a strip of roadway over the pontoons. On this narrow, undulating surface would travel the entire force—men, horses, cannons on their wheeled caissons, ordnance wagons, supply wagons, everything. Colonel Heg, conscious of the honor and responsibility of leading General McCook's entire corps across the river, had been up all night supervising the unloading of the boats and the disposition of his troops.

By the end of the day all of Heg's brigade had crossed the Tennessee, driving off a small force of Confederate pickets in the morning and reaching the summit of Sand Mountain by nightfall. The rest of the corps joined them in a day or two. As they were moving south, word came to them that Chattanooga had fallen to the Yankees—without a shot, it was said. It seemed that General Bragg, the Confederate commander, was on the run. They kept marching south, day after day, until they were thirty miles from their Tennessee River crossing. Then, nearly two weeks after they started, the order came to reverse course; General Bragg was not retreating; he was getting ready for a full-scale battle.

They made better time on the way back, knowing they had to move with all possible speed. By midnight of September 17 Rosecrans' army was bivouacked three or four miles west of Chickamauga Creek, where the Confederates awaited them. On the eighteenth they were moving into position to repel the attack that all expected. Somehow Colonel Heg found time to write a letter to Gunild that day. He warns his wife that the impending battle may be a big one, but he tells her not to feel uneasy for him; he is trusting to his "usual good luck." His faith in his commanding general remains strong; he is sure that Old Rosey will give the Rebels "one of the biggest whippings they ever had." He is concerned, however, for his soldiers; many of them "have marched hard and had rough time for the last two weeks and ought to have a few days rest before fighting." He is concerned as well for Albert Skofstad, who has been so ill that he had had to travel in a buggy the colonel had brought with him. But Albert is better now.

Seth too must have been concerned about his friend's illness and glad that he was better. He may have been in the tent with the colonel, waiting to take the letter to the courier, along with his own letter to Amorette. Many letters to wives must have been written that day.

On the afternoon of September 19, Heg's brigade was fighting for its life in the woods and fields on either side of the Lafayette Road. On the east side thick woods came up almost to the edge of the road; on the west lay two farms. The colonel had led his brigade eastward across the road but their progress slowed once they entered the woods. Moving with difficulty through the tangle of underbrush, peering through the smoke that hung in the trees, the men could barely see their targets. The roar of the guns—the big cannons at a distance and the rifles close at hand—was like nothing these men had ever heard before. It sounded as if the whole world were exploding around them, above them, under their feet. Still they went forward—firing, reloading and firing again—and for a time it seemed that they had the upper hand. But the Rebels pushed harder, their enfilading fire cutting a swath all along Heg's line. The brigade began to falter and fall back. Closer and closer to the road the advancing Rebels drove them. Now the entire brigade fell back to the other side. Bowing to the inevitable, Heg allowed his men a short respite, and then he rallied them. Again they charged; again they were forced back.

That was the pattern that day—advance and withdraw, forward and back—both sides locked in its futile rhythms. James Abernathy, a colonel now, described it in prose that seems to fall into the cadence of a poem: "Reforming the line, we again advanced under a perfect shower of bullets, sometimes driving the enemy and in turn being driven by them, until we had fought the ground over and over again, and almost half of our number lay dead or wounded upon the field."

Late in the day Colonel Heg and two other brigade commanders rallied their men for another charge. They crossed the road again, only to be driven back once more. Heg rode among the men, urging them to fight. He was an easy target. An enemy bullet struck him "in the bowels," as Albert Skofstad reported. Still the colonel remained in the saddle, still the leader of his men, until weakness forced him to relinquish his command to Colonel Martin of the Eighth Kansas. The stretcher-bearers put Colonel Heg in an ambulance, which took him to the field hospital at Crawfish Springs. There the doctors, seeing that little could be done for him, gave him a tent in which to die. Through the night his officers came to pay their respects. Surely his aide-de-camp was one of them.

In the two and three-quarters years that Seth spent in the army, he was in only two battles, Chickamauga and Missionary Ridge. But they were decisive battles, and the brigades in which he served—a different one for each battle—played a crucial role both times. In each case the actions of his brigade prefigured the outcome of the battle. The unhappy story of what happened to Martin's brigade (formerly Heg's) on the second day of Chickamauga has been told and retold, not only on the printed page but on the site itself. Two of the ten stops on the tour of the battlefield at Chickamauga and Chattanooga National Military Park, the stop labeled "Mixup in the Union Command" and the stop called "Confederate Breakthrough," mark scenes from that story. Peter Cozzens' exhaustive history of Chickamauga, *This Terrible Sound,* describes in painful detail the predicament in which "Martin's tiny brigade," as he calls it, found itself.

What began as a simple misunderstanding grew uglier when the factors of anger, resentment and spite entered the equation. First, General Rosecrans received an erroneous report that a gap had developed in his line. Unsure whether the report was correct, he dictated an order that would plug the gap, if one existed. The staff officer who wrote down the order got it wrong; he forgot the "if" clause. The general to whom the order was given knew he should not obey it, for he knew there was no gap. But he was still angry because Rosecrans had publicly rebuked him that same morning for failing to obey orders promptly. And so he determined to obey this new, ill-founded order and pull his entire division out of line, regardless of the consequences. Their departure left a real gap, and the only unit available to fill it was Martin's brigade, reduced by nearly half its members after the fighting of the previous day.

Shortly after noon Colonel Martin, having held the command of the Third Brigade for less than twenty-four hours, was doing his best to carry out an impossible order. With the remnant of his force, he was to fill the space that an entire division had inexplicably vacated. Martin was moving his regiments as fast as he could toward a small barricade at the edge of the woods. Once in position, the men faced an open field owned by a farmer named Brotherton. Martin had got three of the regiments in place—the Eighth Kansas, the Fifteenth Wisconsin and the Twenty-fifth Illinois—but the last one, the Thirty-fifth Illinois, was still filing into position when, as he later reported, "the enemy rose up from the tall weeds in front and advanced on us, four columns deep, pouring in a destructive fire." His men fired back "with terrible effect," Martin wrote proudly. But when masses of Rebel soldiers began climbing over the barricade, while at the same time other Rebels were coming at them from the side, the men of his brigade could take no more, and they fell back in disorder. Farther and farther into the woods they were pushed, back across another field and into another row of woods. At that point, Martin reported, "they were rallied," and they "left the field by order." In other words they did not shamefully flee, they retreated under orders from him.

Cozzens, however, says they were routed. *Routed*—defeated and forced to retire in shameful disorder. Whatever it should be called, the fact remains that the brigade was driven from its position, leaving a wide-open hole in the line, through which the Rebel soldiers came pouring. Other brigades were likewise overcome. Up and down the line panic-stricken soldiers were breaking and running. As the bad news spread, the generals began to leave the field. Rosecrans, it was said, was the first of them to take the road northward to Chattanooga.

Nobody blamed Martin's brigade for failing to hold the line when the crowds of screaming Rebels mounted the barricade and fell upon them. But to those of its members going down the road to Chattanooga in the gathering darkness, the memory of their failure must have been bitter indeed. Nevertheless they could be grateful that they had come through the battle of Chickamauga with their lives. Seth, for his part, might have wondered at his own narrow escape. In the two-day battle he had had two horses shot from under him—a story Amorette would put into the record fourteen years after her husband's death. More grievous still, he had lost the officer with whom he had worked closely, every day, for most of the last three months. Why he had been spared could not be known. That he had been spared was reason enough for gratitude.

A few years ago, on an afternoon in early fall, my husband and I were walking across the Brotherton field. It was shorter from front to back—that is, from the Lafayette Road to the woods at the field's western edge—than I had imagined it would be. Buttercups covered the ground, so many that we could hardly take a step without crushing the shiny yellow petals. When we reached the woods, we turned to our left and walked alongside them. I kept peering into the thicket of trees and shrubs, looking for the Eighth Kansas marker that I had been told was in there. At length I caught sight of a gray stone column, about the color of the smaller tree trunks. I stepped through an opening in the brush and saw that the marker was an obelisk in miniature, with a pyramidal top. On it were the words,

POSITION
OF THE
8TH KANSAS VOL'S
SEPT 20TH 1863
AT 12:00 NOON

Here was where the men of the Eighth Kansas had crouched, behind the barricade, flanked on either side by their comrades of the Wisconsin and Illinois regiments. From here they saw the enemy pushing though the tall weeds, running forward.

Were you here, Grandfather? Were you here at the woods' edge, seeing the columns of charging Rebels come pouring across a field that now looks too small to hold so many? Time has diminished the scale of the stage on which this action was played. The modern eye expects a grander setting. But the reduction in scale does not diminish the import of the scene. If anything, it increases it. The actors loom all the larger for the smallness of their stage, and the magnitude of the action emerges all the more clearly. Here on this shallow field the battle of Chickamauga began to be lost.

Chapter 10

A Home on Union Street

About the time Seth and the rest of McCook's corps were crossing the Tennessee and marching south, which turned out to be a mistake, Mary Rich bought a residential lot in Fayette. It was located near the south end of town, at the corner where Union Street, which ran north and south, intersected with the County Road, which ran east and west. I am assuming that there was a house on the lot and that Mary had bought it as a home for herself, her daughter and her grandson Dell. No house stands there now; the land that was once Mary's now borders a structure at the southern edge of the Upper Iowa University campus. But in 1863 the southern boundary of the campus was Madison Street, which lay two blocks north of the County Road.

The proximity of the university may have been one of the reasons Mary bought the lot. At this time Upper Iowa University had been in existence for six years. It possessed only one building, but it was a beautiful one, already a landmark. Three stories high, built of creamy limestone, College Hall was a simple rectangle in shape. Two pointed cornices ornamented the roofline. A cupola, rising from the center of the roof, housed the bell that was rung nineteen times every school day. On Sundays it called Methodists from the town and the surrounding countryside to a worship service. The use of College Hall as a place of worship was both appropriate and necessary. The statewide Methodist organization sponsored and managed the university, and the Methodist congregation in Fayette had no church building of its own at that time.

Fayette was half-encircled by a small river, the Volga, which in a dry season shrank to the width and depth of a creek. After making its way northward through the fields west of town, the Volga turned right and flowed east, serving as a natural boundary for the north end of Fayette. When it reached open country again it curved to the south and, leaving the town at last, headed southeast on its way to join the Turkey as it flowed toward the Mississippi. Like its river, the town of Fayette was not very big. In those early days it contained no more than seven hundred inhabitants; even counting the university students the population did not reach a thousand. But it was important enough to be visited by "daily stage runs from Independence, Postville and McGregor," as the university catalogue reported. The stage from McGregor would have been the one that brought letters from soldiers serving east of the great river, reopened now from source to mouth by the fall of Vicksburg.

I imagine Mary Rich and Amorette and Dell moving into their new home on Union Street. I see the two women unpacking keepsakes that had traveled with them from the house in Ohio and had lain in trunks and boxes since then. I see

Amorette showing three-year-old Dell where he will sleep in his new house.

A month goes by. The last of the summer heat has gone out of the sun, but it shines all the brighter these clear October days. On a Sunday Mary and Amorette—with Dell between them, holding their hands—are walking up Union Street on their way to the worship service in College Hall. They walk in a glory of golden light. Diffused through the red and yellow leaves of the oaks and maples, it falls on them like a blessing.

Now it is Monday and the stage is expected from McGregor. Amorette leaves Dell with Mary, because the long walk would tire him, and takes the County Road west to Washington Street. From there she walks on to the post office in the hope that a letter from Seth has managed to get out of Chattanooga.

Chapter 11

Under the Mountain's Gaze

When I was about eight years old, it was my job to set the tables for dinner. Our dining room was large enough to accommodate four tables. Mother, Daddy and I ate at one of them, or only Daddy and I when Mother was working a late shift at the nurses' registry. The boarders ate at the others. Mary, the girl who worked for us (we never called her the maid), ate in the kitchen so that she could serve us the food she had cooked—her pot roasts, her waldorf salads, her pineapple upside-down cakes. Nineteen years old, educated in domestic science in high school, she was very proud of her cooking, and we were proud of her.

At five o'clock in the afternoon I would come in from playing and go to the kitchen to help Mary. The silverware was kept in a cupboard just inside the kitchen door. I would pull open the wide, heavy drawer and count out enough knives, forks and spoons for the family and as many others as were staying with us at the time. The silverware, which my parents had bought when they were married, was Community Plate, quite plain, with only a small decorative border around the handles. There was one spoon in the drawer, however, which I never used because it did not go with the rest of the pieces. It was entirely different, slender where they were broad, ornamented where they were plain. In the bowl of the spoon a mountain rose into a silver sky, literally rose—I could run my finger over its rocky sides. Its name was engraved on the face of the handle. "Lookout Mountain," it said.

I thought it must have been a souvenir from someone's vacation trip, a trip that must have been taken earlier than I could remember. I had never seen Lookout Mountain, or any mountain for that matter. Someone in the family must have gone there and brought home this spoon, I thought, but I did not know who it was or why he or she went there.

Two hours after Colonel Martin's weary men came into Chattanooga, they were put to work digging rifle pits. When those were done, they spent the rest of the day and part of the night building heavy breastworks to deflect enemy fire. Chattanooga, once a trophy of Union victory, was now a city under siege, backed up against the looping Tennessee River. The best General Rosecrans could do was draw a tight line around it and fortify it, for the Confederates had his army nearly surrounded. Not only did they control the flat plain outside the Union line, they also controlled the heights: Missionary Ridge, a six-mile-long chain of high steep hills that lay about three miles east of the city, and Lookout Mountain, which rose eighteen hundred feet above the river.

The campsite of Colonel Martin's brigade faced the mountain. His men were on the same side of the Tennessee as it was, but because of the great bend the river made they looked at the soaring cliff from across the water. At first, Lookout seemed to pose a deadly threat. Every day the big guns on its summit would flash and boom as they attempted to fire upon the Union camp. But few hits were scored, and the men down below began to treat the daily displays as more of a joke than a danger. The dwindling food supply, however, was no joke. The Confederates had Rosecrans' army almost cut off from its supply base. For the men, real hunger began to set in; for the animals, it was starvation. Rations were cut down to one-half, then to one- third, so that the men were subsisting on three or four good-sized crackers per day, with the occasional addition of a piece of unpalatable meat. Officers, whose higher pay might have allowed them to buy more food than the enlisted men received, were forbidden to do so.

All the while the mountain seemed to be looking on. This was not a mere fancy—up on its rocky ledges a Confederate with a good pair of field glasses could see the tents of the Union camp, could watch the Union soldiers, doll-like in size, going about the task of survival. For those soldiers, laboring and suffering down on the river-plain, it would not have required too much imagination to suppose that the mountain itself was watching them. On cloudy days, which were many, it was a spectral presence, its intentions impossible to decipher. At the end of a clear day the mountain was gray-blue against a sky of dusky orange. Then it appeared more benevolent; then one might infer a message of comfort. But kindly or threatening, concerned or aloof, it was always there; it never averted its gaze,

In October Martin's brigade lost its identity when it was subsumed by Brigadier General August Willich's First Brigade. Colonel Martin returned to the command of the Eighth Kansas, and all the regiments of the disbanded brigade moved a mile and a half to the east, where Willich's force was encamped. Now they faced Missionary Ridge and, closer at hand, a round hill—a line of Confederate breastworks running over it—that bore the homely name of Orchard Knob.

These changes in leadership were part of a much larger reorganization. General Rosecrans was removed from his position as commander of the Army of the Cumberland. General Grant, newly arrived to take charge of the entire operation in eastern Tennessee, had replaced him with General Thomas. The presence of Grant, who had a reputation for getting things done, gave hope that the Union forces might be able to break out of the besieged city. Nevertheless the soldiers were still hungry and the horses and mules were still dying. Then one day, when October was almost over, came the news that "the cracker line was open." In a daring operation a hand-picked group of men had defeated the Rebels who stood between them and their supply base and had opened up a short, good road to food. Early in November rations went up from one-third per day per man to two-thirds.

For Seth, the reorganization meant a new job. Now he was ordnance officer for General Willich, responsible for the weapons and ammunition to be used in the fighting. Among other duties, he would supervise the packing of the wooden chests that held the several types of ammunition for the cannons. There was the round cannonball of solid iron—the "bowling ball" of army ordnance—favored for use upon a column of troops, where it could quite literally bowl them over. There was the shell, cylindrical, hollow, filled with explosive, depending for its lethal effect upon bursting. Most often it burst during its screaming flight through the air, and in jagged fragments tore into the bodies of its victims. The other two kinds of ammunition used by the cannons were case and canister; similar in shape to a shell, they were filled with small balls of iron or lead. They differed from the shell in that it was the contents, rather than the outer casing, that were designed to kill or maim.

Once again, as in the summer just past, Seth was working for a man who had come to the United States as an immigrant. August Willich had been an officer in the Prussian army in the 1840s, but, too much of a democrat to be comfortable serving under the Kaiser, he joined the forces of the revolution. When the revolution failed, he was forced to flee Germany, and in 1858 he came to the United States. Upon the outbreak of the Civil War Willich, true to his democratic principles, enlisted in the Union army as a private, but his talents were quickly recognized and he was soon promoted. He was known to be a highly competent and scrupulously honest officer.

On November 23, a bright and sunny day, nearly twenty-five thousand members of General Thomas's Army of the Cumberland were taking positions on the plain west of Orchard Knob, preparing to make an attack on the enemy's line. At noon the order came: Willich's brigade, with the Eighth Kansas in the lead, was to advance upon Orchard Knob at the sound of the bugle. The order stipulated further that the brigade was "to take Orchard Knob and the line of breastworks running over it. . . if possible." With no time to spare, the men of the Eighth Kansas were issued a fresh supply of ammunition. Issuing that fresh supply would have been Seth's job.

The bugle sounded and the advance began. Willich took Orchard Knob that afternoon with remarkable speed, which he credited in part to the "splendid advance" of the Eighth Kansas. His brigade lost only fourteen men (four killed, ten wounded), and Seth was not among them.

The next day the brigade, newly positioned on Orchard Knob, could tell that a battle was taking place on Lookout

Mountain. With the naked eye they could see the smoke rising from the big guns, and with field glasses they could make out the tiny figures of men, climbing upward. But between the intermittent rain showers and the clouds that hung over the summit, the spectators were hard put to know who was winning. Somebody reported seeing the stars and stripes, but then the clouds came back and the onlookers were not sure. Morning came before they learned that the Union forces had driven off the Rebels; Lookout Mountain was now in Federal hands. That victory left one more height to capture, Missionary Ridge. No one expected it to be as easy as the other two had been.

It was chance and nothing more, but at Missionary Ridge, in the second of the two major battles in which Seth ever participated, his brigade once again played a pivotal role. As at Chickamauga, the brigade's action prefigured the outcome of the battle. And again the outcome turned upon a misinterpretation of an order from the commanding general.

Early on the afternoon of November 25, two days after the first great assembly of forces on the plain, another one took place. Now Orchard Knob lay behind Thomas's army, and Missionary Ridge, bristling with cannon on its summit and defended by enemy rifle pits along its base, was closer than it had been before. Between the assembled army and the Ridge was, first, a belt of woods that offered the attackers some protection from enemy fire. But then they would have to break out into an open field, cross it, ascend a strip of rising ground, and at its farther edge overpower the occupants of the rifle pits. Beyond the rifle pits stretched a plateau, about a hundred yards wide, and then the slope reared steeply upward to a height of almost five hundred feet.

Willich had drawn up his brigade in two lines: four regiments in the first line; another four, including the Eighth Kansas, in the second, with the Fifteenth Wisconsin placed behind as a reserve. He, along with the other commanders, was waiting for the six successive cannon shots that would signal the start of the operation. General Grant's order, as Willich understood it, was to "advance" upon hearing the sixth shot. That was all. Not advance and stop somewhere short of the top of the Ridge. Just advance. But the order as Grant issued it called for the Army of the Cumberland to "'advance and take the enemy's first line of rifle pits.'" There was no instruction to advance farther.

The series of cannon shots began. After the sixth boom the regimental commanders called out the order to march, and the army began to move forward. Here, from W. S. Burke's history of the Eighth Kansas, is what it was moving into: "Through the branches of the leafless trees, we saw a bright flame leap out and a dull grey smoke curl up all along the summit of the Ridge; a crash like a thousand thunder claps greeted us; solid shot went screaming through the timber, and hurtling shells exploded above and around the lines, sending their scattered fragments shrieking through the air like a legion of devils."

In quick time, not stopping to fire their rifles, Willich's men crossed the open field. When they reached the rising ground, where shell and case were falling thick and fast, they double-quicked toward the enemy's rifle pits. The Rebels in the pits stood their ground at first, firing their guns at the oncoming attackers. Still the Yankees kept advancing, until they had gained possession of the rifle pits themselves. At that, some of the Rebel defenders turned and began running toward the safer haven of the Ridge, while others threw down their arms in surrender and leaped over to the Union side of the entrenchments. Seeing that there was not room for attackers and prisoners in the same narrow shelter, Adjutant Sol Washer of the Eighth Kansas dragged one of the prisoners out into the open and ordered them all to the rear of the advancing army. As he did so, he taunted them: "'You have been trying to get there long enough. . . now charge on Chattanooga!'"

In point of fact, the prisoners were better off to head to the rear than to remain where they were. By now the shot and shell were raining down incessantly upon the rifle pits. Here was where General Grant had intended the army to stop; he did not think it feasible for them to climb the Ridge under fire. But the commanders responsible for carrying out Grant's order—both those who had received it correctly and those who had not—could see that it was suicidal to remain here.

General Willich concluded at once that the only course was to go forward. Accordingly he sent members of his staff, including Seth, to convey his order to the four regiments of the first line. The report Willich later wrote tells how he parceled out the assignments: "My adjutant, Captain Schmitt, was already at the extreme left. I sent my aide, Lieutenant McGrath, and ordnance officer, Lieutenant Foot (who on this occasion was wounded by a shell), to different regiments, I myself, with my inspector Lieutenant Green, went to the Eighth Kansas, and the command forward was soon heard all along the lines." Captain Schmitt was to give the order to the regiment on the left, where he was. Lieutenants McGrath and Foot were to do the same for those in the center and on the right.

I try to visualize what Seth did after receiving the order. Already mounted, he must have spurred his horse and ridden off toward the regiment he had been assigned. In such a storm of shot, shell and case, it seems that the best protection would be to ride fast, as you would run through a rainstorm, hoping to dodge the pelting raindrops. So he may have ridden fast, but the shell found him all the same.

All along the base of the Ridge the soldiers of Thomas's army were advancing—running across the rising ground, climbing up the rugged slopes, some under orders, some spontaneously. General Willich's brigade was making better time than most of the others. As his men climbed higher, the danger from above lessened, for the cannons were now firing over their heads. The men in the center and right of Willich's line, along with men from General Hazen's and General Beatty's brigades, who in the confusion of the ascent had mingled with them, reached the crest first. There in the face of heavy fire they drove off the Rebel defenders. The honor of being the first to plant its regimental colors on the summit was shared by one of Willich's regiments and one of Hazen's, but within a few moments the flags of all Willich's regiments were flying on the crest. Then more and more Yankees, cheering, were chasing the Rebels down the far side of Missionary Ridge. It was a proud and joyous day for Willich's brigade and for the entire Army of the Cumberland.

In this scene of jubilation Seth takes no part. He lies on the ground, somewhere near the rifle pits, until the stretcher-bearers find him and carry him off the field.

It was probably my father who bought the Lookout Mountain spoon. He might have picked it up in the 1890s at a souvenir shop in the newly established Chickamauga and Chattanooga National Military Park. The site was officially dedicated in 1895, "with impressive ceremonies," according to the Park Service brochure. He might even have gone to the ceremonies, as the son of a veteran of Chickamauga and Missionary Ridge. Perhaps he bought the spoon to give to his mother, thinking it would serve as another token of Seth's war service, along with the photograph and the shoulder straps and the bundle of letters. This small hoard was almost all Amorette had left of her husband's presence in her life. But the spoon does not quite fit in with the other things. Seth had handled the photograph, worn the shoulder straps, written the letters. The spoon, for all its symbolic intent, had never felt his touch.

Chapter 12

Convalescence

After the battle it was Colonel Martin's responsibility to draw up a list of casualties for the Eighth Kansas: so many dead, so many wounded, so many missing. Among the wounded he included Seth, who still belonged to Martin's regiment even though he was serving General Willich at the time he was injured. Martin reported that Seth was wounded severely in the thigh. These three pieces of information—he was wounded, the wound was severe, it was located in his thigh—constitute the only official medical record of Seth's case that the National Archives possesses. Beyond these three facts I have had to look at other sources to discover what kind of care and treatment he probably received.

We can be quite sure that the stretcher-bearers reached him quickly and took him by ambulance to a place where he could be treated. Arrangements for the care of the wounded were much better after Missionary Ridge than they had been for Chickamauga, where many wounded men had lain on the battlefield all night, untended. By two o'clock in the morning after Missionary Ridge all the wounded men had been taken to shelter. I am assuming that Seth, with his severe wound, would have been treated in one of the hospitals which the army doctors, knowing a battle was inevitable, had set up in advance in Chattanooga.

His initial treatment would have gone something like this: A doctor would examine the wound. if he saw shell fragments in it, he would try to remove them, perhaps with an instrument (unsterilized), perhaps with his fingers (unwashed). When this ordeal was over, Seth was probably taken to a cot where he could rest and recover a portion of his strength. In a day or so his wound probably became infected. The onset of infection, twenty-four to forty-eight hours after the wounding, was such a common occurrence that doctors regarded it as a normal stage in the recovery process. They called it the "irritative stage."

The doctors in Chattanooga were well aware of the danger of uncontrolled infections, but they did not know what caused them. One popular theory was that it was bad air. Realizing that the churches, schools, warehouses and private homes that were serving as hospitals lacked adequate ventilation, the doctors preferred to house their patients in tents, where fresh air could circulate. Unfortunately they did not have enough tents to accommodate all the wounded men. Dr. W. W. Blair, medical director for the Third Division (Seth's division) assigned his wounded to tents until the supply ran out and then sent the remaining men to the general hospital. Seth probably spent a few days in one of these facilities, recovering a portion of his strength.

But then he would have had to leave. Officers were not permitted to stay for any length of time in Army hospitals, which were intended only for enlisted men. With their higher pay, officers did not qualify for the free care the Army hospitals provided. Some of them went home for care, some went to private hospitals and some stayed in their quarters, where their orderlies acted as nurses. This last choice is the one Seth probably made. The muster roll for the month of December 1863 shows him still on detached service with General Willich's brigade. I imagine he spent all or most of that month resting in the tent where he had slept the night before Missionary Ridge. During his convalescence his orderly would have brought him his meals and dressed his wound.

Of course I am using the word ironically. I do not think Seth was really convalescing; I believe that bacteria had entered his wounded body, transmitted very likely by the doctors' hands or instruments. I do not blame the doctors. From what I can tell, the medical men who treated the wounded at Chattanooga were hard-working and conscientious, truly concerned to give their patients the best possible care. But they did not know about germs. That knowledge was on its way; Ignaz Semmelweis and Louis Pasteur had already presented evidence that would lead to the development of the germ theory of disease. But the new knowledge—not widely accepted, even in Europe—did not come soon enough to help the Civil War soldiers.

It is clear to me that Seth developed an infection soon after he was wounded and that the infection resisted any efforts to cure it. Later chapters in this book will provide details. For the present, two examples of soldiers with similar case histories offer context for this part of Seth's story:

Shelby Foote recounts the experience of Brigadier General Carnot Posey, "struck in the thigh by a fragment of shell. . . . the wound, though ugly, was not thought to be grave, but infection set in and he died one month later." A case study in the records of the United States Sanitary Commission describes the lengthy convalescence of a soldier who was wounded at Missionary Ridge. The soldier's wound became infected, and he was treated at the hospital for two months. When he returned to duty, he was transferred to a more distant post. But his wound was still inflamed, an abscess formed, and he returned to the hospital for more treatment. By May 1, six months after Missionary Ridge, he was reported nearly recovered: "wounds in good condition, and general health improving."

There are parallels between this soldier's case and Seth's, but to me, Seth's granddaughter, they only make the differences more poignant. For Seth seems to have been considered fit for duty in little more than a month after the battle. The Company A muster roll shows that he was present with the company during the months of January and February 1864. Considering the examples I have cited above, and knowing the way his story ends, I am convinced that when he returned to duty his wound was still infected.

At the time Seth rejoined his comrades, Company A, along with the rest of the Eighth Kansas, was encamped at Strawberry Plains, Tennessee, about ten miles northeast of Knoxville. On the second day of the new year the members of the regiment were asked to re-enlist for another three-year term. Two days later four-fifths of them did so. The mood of those who signed on again was different from what it had been back in 1861, as the regimental historian later reported. No longer were they unconscious of the terrors and hardships of a soldier's life, yet they were willing to endure them for three more years if necessary.

Not all the officers who figure in Seth's story chose to fight on. Lieutenant Colonel James Abernathy—the man who had organized Company A, the officer who had taken command of the Eighth when Colonel Heg was carried off the field and Colonel Martin was called to replace him, the writer whose description of the ebb and flow of the fighting that day reads like a poem—Abernathy resigned from the army in early November, before the battle of Missionary Ridge. The brief notice in the regimental history provides no explanation for his decision.

But Sol Washer re-enlisted. On the day of Missionary Ridge he was still not recovered from the wound he had received at Chickamauga. Furthermore, he was suffering from a severely sprained ankle, which made it difficult for him to walk. Nevertheless, "he mounted his horse," Colonel Martin reported, "and rode through the whole battle, always foremost in danger." It is difficult to see how his horse could have carried him all the way to the top of the Ridge, steep as it was, but no matter. I do not question his fortitude. On the contrary, my point is that his injuries did not discourage him from re-enlisting.

Nor did the injury Seth suffered discourage him. The record of re-enlisted veterans of the Eighth Kansas includes his name. Seeing it, I am impelled to ask, why? Why didn't he go home as soon as he could and perhaps recover?—a question Amorette must have asked herself more than once in the years to come. One reason may have been money. Though he, as an officer, probably did not qualify for the re-enlistment bounty the War Department was offering to enlisted men, his regular military pay would have provided some incentive.

Nevertheless, it is hard to believe that money alone would have made him willing to face again those murderous barrages of shot and shell that he had known before. I think he must also have felt the twin impulses of "cause and comrade" which, as James McPherson has shown, drove so many Union soldiers to go on fighting despite their bitter past experiences. Over half of the Union men whose terms expired, as Seth's did, in 1864 chose to re-enlist. After studying the letters of a representative group of these soldiers, McPherson has concluded that belief in the absolute necessity of preserving the Union was the most powerful force behind their decision. Next came loyalty to their comrades in arms; having lived and suffered beside their friends for so long, they did not want to abandon them now.

Without Seth's letters I have no proof that he felt like these others. But the glimpses of him that existing records do permit—his service to Colonel Heg and General Willich, for example, his friendship with Albert Skofstad, even the convivial poker game with the cavalrymen and the gentleman gambler—make me think that duty to country and loyalty to comrades carried more weight with Seth than money.

The weather in East Tennessee in January of 1864 did nothing to aid the recovery of a wounded man. Rain, sleet and snow, sometimes one, sometimes the other, fell day after miserable day. The Eighth Kansas marched to a town named Dandridge to repel an expected attack, they several times lined up in battle formation, but the attack never came. Dr. Blair, who had accompanied his division to East Tennessee, complained that "a very considerable number of men" suffered seriously from "exposure during this extremely fatiguing march." But little by little the regiment's situation improved. The weather warmed up, for one thing. The Eighth returned to Knoxville, where they were promised a furlough. By Valentine's Day they were on their way back to Kansas.

Seth's furlough order, dated March 1, gives thirty-five days of leave to all the men of the Eighth Kansas who re-enlisted, calling them "Veteran Volunteers." Singling out Seth and the other officers by name, it directs them to rejoin the regiment at Fort Leavenworth on April 5. It asserts, further, that during their absence the officers "will procure for the Regiment all recruits possible to obtain."

Chapter 13

Dying at Home

Now the veterans day by day die out.

— August Willich

To get to the land that Moses Foot once farmed, you take a gravel road west out of Farmersburg. You go up a long, not very steep hill, passing a few old farmhouses, and when you are almost at the top of it, you turn right. You drive north half a mile or so, past one or two newer-looking farmhouses, and then you turn right again and go down a lane with fields on either side. At the end of it you come to a farmyard where some cattle are penned, for this is still a working farm. Just behind the cattle-pen is a long shed, open on one side, that holds feed for the cattle, and diagonally opposite the shed is a red and white barn. Both these buildings are fairly new and in good repair. But on the other side of the farmyard, up a slight rise, stands an extremely old house. As far as anyone can be sure, it has not been lived in since the nineteen twenties. Most recently it has been used as a machine shed.

The house is a two-story box, of a gray-brown color, the color clapboard siding gets when it has not been painted for a great many years. The front is as plain as a house can be, except for a small enclosed porch that frames the door. The porch is roofed by an inverted V. Above it is another, larger inverted V, this one not a roof but only a stain on the face of the house, its point up near the eaves, its two arms extending downward and outward between the upstairs windows. The stain shows where a roof used to be.

According to county records, Moses and Mary Foot owned this land from 1855 to 1865. It was located in Section 12 of Wagner Township, Clayton County, Iowa. The nearest town of any consequence was Farmersburg, which is why people who lived in Wagner Township were inclined to give Farmersburg as their place of residence. This piece of land—about half of it wooded in the 1860s and the other half open fields—was the home place to which Seth returned in the spring of 1864. But was this the house in which his father's family lived? The family that owned the farm before its present owner say they do not believe it is old enough. The really old houses in the neighborhood, they say, have hand-hewn logs in their internal structure; this one does not.

Still there is that stain that marks the front of this house. It must have been made by another structure, now gone. Possibly the existing house was itself an addition to that older, smaller house, later demolished. Perhaps, if we are not now looking at the old Foot farmhouse, we are seeing its traces.

For a couple in late middle age (they were both fifty-one that spring) Moses and Mary Foot had heavy responsibilities. With their older children grown and gone they had their second family to raise, thirteen-year-old Alfred Foot and his ten-year-old sister Mary Elizabeth—still in school, too young to be of great help with the work of a farm. And farm labor was hard to come by in this fourth spring of the war. "Our boys have gone to war," the *Clayton County Journal* wrote, "and we want others in their places." To find those others to plant and harvest their crops, the farmers of the region had begun to look for workers who were unlikely to be drafted, that is, newly arrived immigrants from other countries. George Wakeman, the same George Wakeman who had gone off to the mines with Seth and Gurdon and the others, was now heading a drive to recruit foreign laborers for work in the fields of Clayton County. How Moses coped with the labor shortage we cannot tell, but it is safe to say that his life was not easy that spring.

Now that Seth was coming home, weak and sick, Moses had a new source of anxiety—and increasingly a source of expense as well. Even before Seth had taken the steamboat down the Missouri to St. Louis, and another steamboat up the Mississippi to McGregor, and had limped ashore, flushed and haggard, Moses had begun keeping a running account of the expenditures he was making on his son's behalf. On December 1, just six days after Seth was wounded, Moses paid ten dollars to Amorette. He made another such payment on January 5 and still another on February 20. These payments evidently substituted for an allowance that Seth had been sending her before he was wounded. On that same February day Moses, anticipating the extra breakfasts that would have to be cooked, entered in Seth's account the purchase of seven pounds of butter and eight pounds of sausage.

Seth arrived at his father's house in early March, unable, it appears, to travel on to Fayette where his wife and child were. On March 17 Amorette came to Moses' house to stay. From then on, nearly to the end, she was probably the one who gave Seth most of his nursing care. Mary Foot would have been too busy with her housekeeping; Moses would have been too busy with his farm work and with the trips to town that Seth's illness made necessary. Furthermore, it was Amorette's place, as Seth's wife, to tend to her husband, and to have a voice in decisions regarding his care. Four days after she arrived, a doctor came for the first time to attend Seth. It sounds as if Amorette, seeing Seth's condition, had urged that the doctor be called.

When Dr. Horace Hamilton made his initial visit, he left Seth some medicines, the cost of which he carefully recorded. He returned twice more in March, and in April he began coming nearly every day, always with the medicines. In addition he must have prescribed alcohol, for Moses bought a pint of brandy on April 5 and another on April 8, noting the cost on Seth's account. Alcohol was the "sovereign remedy" of the Civil War, as medical historian George Worthington Adams remarks. It may have pained Moses, with his strict religious principles, to supply his son with alcohol, but he could take comfort in the belief that it was being used only for medicinal purposes.

The day Moses bought the first bottle of brandy was noteworthy in another respect. It was the day when Seth's furlough ended, the day he was due back at Fort Leavenworth. But in order to arrive on time, he would have had to leave several days earlier, and by the beginning of April he was obviously too ill to do so. On April 2, therefore, Dr. Hamilton wrote to Seth's regiment, explaining the reason for his absence: "I hereby certify that I have this day examined Seth Foot a Lieutenant of Company A, 8th Regiment Kansas volunteer infantry and find him suffering from opthalmia and acute Rheumatism - confined to the bed - and in my opinion he will not be fit for military duty within twenty days."

Ophthalmia, a hard word to spell (and Dr. Hamilton misspelled it), means "inflammation of the eyes." Rheumatism is characterized by pain and stiffness of the back and extremities. These two conditions, though uncomfortable, are not necessarily disabling. Seth was suffering a worse condition, an infected wound that would not heal, and Dr. Hamilton knew it but did not say so.

When Sol Washer, adjutant for the 8th Kansas, received Dr. Hamilton's certificate, he sent Seth specific instructions for any subsequent certificates that might need to be written:

> Lt. Seth Foote
>
> Your pr Dr. Hamilton received also your certificate which I beleive covers the case You should provide yourself with triplicate certificates one ford to Adt-Genl Washington one to Regtl HdQrs retaining the other yourself. these will have to be forwarded Evry 20 days during your Absence.
>
> Hoping that you may soon be able to re-join the Regt I Remain
>
> Respectf
Your Obt Svt
Sol R Washer
Adjt
>
> P S The Regt is under marching orders for Tenn Will Leave on first Boat

The ending of Washer's letter, after the bureaucratic details have been disposed of, speaks directly to the Civil War soldier's feeling of attachment to his comrades, that strong bond which, for better or worse, draws him back to them. Reading the brown and faded words, I grieve for my grandfather, unable to rise from his bed and go where his loyalties would pull him. For a time at least, he seemed to hold on to the hope that he would return to the regiment, and he asked to have his watch repaired. Someone named A. French repaired it for him, at a cost of $3.50, which Moses added to Seth's account.

The twenty-second of April was the date when the second medical certificate was due to be sent, but since Sol Washer did not write his letter of instructions until the nineteenth, it was impossible for Dr. Hamilton to get the second certificate in on time. When he did write it, he altered his diagnosis somewhat, dropping *ophthalmia* and describing Seth's symptoms as "Rheumatism and extreme debility." Then the doctor had to take the certificate to be notarized, and it was April 29 before it was ready to be mailed to Leavenworth. Since it would have had to go to Kansas by boat, down one river and up the other, it could not have arrived until the first week in May at best. When it did arrive, the change in diagnosis might have given rise to the suspicion that Lieutenant Foot was malingering, for "extreme debility" could cover all sorts of vague complaints.

No matter what Dr. Hamilton told the Army, there can be no doubt that he recognized the true nature of Seth's illness. Seth had pyemia, "the dreaded pyemia," as Adams describes it. The name, which translates as "pus in the blood," was commonly used by doctors of the time to describe a blood-borne infection that developed as a consequence of a wound and that was accompanied by one or more abscesses. It was almost always fatal; of those stricken, ninety-eight per cent died.

Why did Dr. Hamilton withhold the name of Seth's illness from the Army in the spring of 1864? I suggest that he was motivated by wartime patriotism, that he, as a man of some influence in Clayton County, was manifesting the prevailing reticence to speak openly of the potentially fatal consequences of military service. The Army needed more soldiers; the President had recently issued a call for 200,000 more men. Every

Congressional district in the Union had its quota of volunteers to fill, and any district that fell short would be subject to a draft. The governor of Iowa was doing all he could to encourage volunteering, even going so far as to forbid any person to cross the Missouri River, Iowa's western boundary, for the purpose of avoiding the draft.

In Clayton County that spring, the *Journal* was making its own contribution to the recruiting drive by issuing sanguine accounts of local soldiers who were home on furlough. Early in March it printed a cheerful news item about a certain Captain E. A. Crary: "This gallant and brave commander of Co. E, 9th Iowa Infantry Regiment is at home on furlough, having, with the balance of the regiment, re-enlisted and come home. The Captain looks heartier than we ever saw him before and it seems the longer he stays in Uncle Sam's service, the better the uniform becomes him." A month later the *Journal* welcomed two more furloughed soldiers, both of them sergeants, with the same formula of determined optimism.

Clearly, the spring of 1864 was not a good time to let it be known that one Clayton County soldier was desperately ill of a wound he received in battle. Seth, home on furlough at exactly the same time as the three others, was never mentioned in the *Journal*. And Dr. Hamilton, anxious perhaps to keep the frightening name of Seth's illness from the ears of the community (for the notary would have had to know, and might have talked) told the Army some of Seth's symptoms but did not reveal their cause.

In the month of April Dr. Hamilton made twenty-three visits to his patient, always bringing medicine though rarely specifying what kind of medicine he brought. On April 23, however, he recorded two ounces of camphor and a pint of brandy. To Amorette must have fallen the task of giving Seth his doses of brandy—all that brandy, nearly half a gallon was bought for him in just under a month—even though her Methodist beliefs and her observation of her husband's condition may have made her skeptical about its usefulness. After the war was over, doctors more and more began to question the efficacy of alcohol as a remedy for acute infection, and twentieth-century medicine came to see it as injurious in such cases, since it caused an already weakened patient to become weaker.

Better perhaps than anyone else, Amorette knew the state of Seth's disease-ravaged body. If, as seems likely, she was the wound dresser, at least on the days when the doctor did not come, she would have had to confront the fetid abscess that was draining her husband's life away. In late April, probably because Amorette had become exhausted, Moses hired two attendants to come to the house and care for Seth. They drove to the farm and stayed nineteen days. Their help must have given Amorette some welcome relief, but their presence caused more expense for Moses since, in addition to their wages, both they and their team of horses had to be boarded.

On the first of May Dr. Hamilton brought a bottle of liniment, intended no doubt to ease the pain in Seth's rheumatic limbs. That same day the doctor brought another half pint of brandy, the last that would be provided for Seth. He made four more visits to his patient, but during the last four days of Seth's life he did not come at all. Possibly the family, despairing of a cure and anxious about the mounting fees, dismissed him. Probably it was clear by that time that all hope was gone. Probably Seth lay in a coma, often the final stage of an overwhelming infection like his.

On May 14 Seth died. He was buried in the Farmersburg-Wagner Cemetery, established by the Wagner Township Cemetery Association, of which Moses was a founding member. The burial ground lay along the top of a gentle hill a few miles south of the town of Farmersburg. From Seth's grave one could see Moses' land on the horizon, at the crest of another long, gentle slope. A gravestone was erected; it is still standing, its inscription still legible:

> LIEUT
> SETH FOOT
> CO A
> 8 KANS INF

Seth was thirty years old when he died. Amorette, at twenty-four, was now a widow. She remained at the home of her parents-in-law for five more days, long enough for the funeral and the burial, and then returned to her mother and young son in Fayette.

It is hard to avoid the conclusion that Seth was not a very good husband and father—chronically in debt, absent on a foolhardy expedition when his child was born, unable or unwilling to return to his family when the mining venture ended. Yet Amorette always honored him. Fourteen years after his death, when she contributed autobiographical facts for the Fayette County Directory, she had more to say about her late husband's military career than about her own life. Knowing little about the Army's chain of command, she exaggerated his place in the military hierarchy, asserting that he had been aide-de-camp to both General Rosecrans (commander of the Army of the Cumberland) and General McCook (commander of the Twentieth Army Corps). He never was, of course.

But he did carry the order that sent one of General Willich's regiments surging up the slopes of Missionary Ridge—an action that would ultimately cost Seth his life. He helped his comrades to victory in that battle, and their victory helped their country win a war which, though unconscionably brutal, was fought in a just cause. We need to remember this about Seth Foot.

PART FOUR

Widow and Son

Chapter 14
Settling Up

After the suffering and loss it is necessary to deal with what might be called the business of death. I turn over once again the pieces of paper—the will, the legal forms, the bills and receipts—all of them copied from Seth's file in the county courthouse, and a play, distilled from these seemingly lifeless documents, begins to take shape in my mind.

THE CAST OF CHARACTERS:

Seth: a presence felt though never seen. His wishes are set down on a piece of blue paper, blotted and smudged.

Moses: He plays two roles. First, he is the executor of his son's will. Second, and equally important, he is the principal debtor—his note for eight hundred dollars is the only usable asset in Seth's estate.

Philander Austin: the principal creditor. He is the mysterious outsider, the one with the power, the one who will exert the strongest influence on the outcome of the action.

Amorette and Dell: If the claim of affection were the deciding factor, their claim would be the strongest. But it is not the deciding factor.

Mary Rich: Quite late in the play she steps to the center of the stage and causes a disturbance.

Angello Rich: He enters even later, in the role of defender and arbitrator.

The appraisal team: They assess the value of Seth's personal property and try to protect the interests of Amorette and Dell.

Lesser claimants: Dr. Hamilton and A. G. Cloudman put in brief appearances—the one to provide information; the other, comic relief.

Presiding judge: Alvah C. Rogers

The plot

A well known one. Several interested parties contend for what each of them considers to be his or her fair share of a deceased person's estate. To say more at this point would give away the climax and the dénouement.

The setting

A farmhouse in Wagner Township, a home on Union Street in Fayette, a courtroom in the county courthouse at Elkader.

The action

When the play opens, Moses is setting out for the courthouse, carrying Seth's will. His son has been dead only four days. The promptness with which Moses undertakes his task says something about the kind of man he is. Almost certainly the ordeal of the past two months—keeping his home and farm running while watching his son die—has taxed his strength, but he has promised Seth and the court that he would be the executor of the will. Allowing himself little time for rest or for grief, he makes his way to Elkader to take up his new duty.

He leaves the will at the courthouse that day. Two weeks later he returns to hear it read in Judge Rogers' court. Its elevated language, conventional as it is, infuses the simple document with dignity and importance: "Know all men by these presents that I Seth Foot of Clayton County Iowa do duly make and declare. . ." The voice reads on, announcing as if it were the voice of Seth himself that he gives, devises and bequeaths one-third of his estate to his beloved wife Amorette and two-thirds to his beloved son Dellizon. That is all, except for the appointment of Moses to execute the intentions expressed in the will. The reading cannot have lasted more than a few minutes, but the task that lies ahead will occupy Moses for over two years.

Moses has run an advertisement in the *Clayton County Journal* calling for debtors to "make immediate settlement" and for creditors to present their claims. First to respond is Philander Austin, Seth's former business partner. We see Moses at his desk in a corner of the farmhouse kitchen, opening Austin's letter. Without hesitation he draws upon the eight hundred dollars he has just deposited in Seth's account and pays Austin one hundred and fifty dollars. His action makes it clear that if he ever had any intention of sheltering this money from public notice, for Amorette's benefit, he has decided against it.

Amorette, however, has another protector, the State of Iowa. Iowa law required that when a married man died a sum adequate to support the widow and child for a year should be deducted from the estate before any other disbursements were made. The appraisal team, consisting of Moses' neighbors Isaac Russell and Samuel Little, as well as a man named James McAlvin, has already arrived at a figure of two hundred dollars for Amorette and Dell, and in fact Amorette has been given about a third of it. But she is not allowed to keep it. Moses requires her to turn it over to him, as executor, and he puts it in the "set-aside" fund, to be paid to her at a later date.

The team of appraisers comes to the Foot farmhouse. They inspect all the articles laid out for them and assess the value of each one. The list they draw up is not very long. One pair of boots, one hat, one pair of pants, one coat, one vest, one shirt, some drawers, some stockings—all that was needed to outfit a soldier. Besides the clothing, Seth left behind a "glass" (looking-glass? magnifying glass?), a pen, a watch and a trunk. The appraisers give these possessions a valuation of sixty-three dollars. In actual cash, of course, the total comes to almost nothing.

Having satisfied Austin for the time being, Moses turns his attention to local creditors. Dr. Hamilton has charged the estate $1.50 for each house call. When he adds the cost of the medicine and brandy, the total comes to seventy-five dollars. A. G. Cloudman's claim is much smaller and he has waited much longer to be paid. In May of 1859 Cloudman pastured Seth's cow and horse for him and rented him a pair of oxen. Interest accrued since then has increased the charge to $11.28, but Cloudman is not claiming the entire sum. He has accepted in partial payment a pair of Stoga boots, as he describes them, and has taken three dollars off the bill, so that he is now asking for only $8.28.

He signs his claim in a burst of Old Testament fervor: "I A G Cloudman do solomonly Swear that the said above account is just and corect so help you God." Moses pays both claims in full.

Several months have passed. Moses is again sitting at his desk. This time he is drawing up a claim of his own. He is asking payment for out-of-pocket expenses he incurred on Seth's behalf and for services he rendered to his son and daughter-in-law during Seth's illness. Here the accounting he kept in the winter and spring of the preceding year proves useful. He notes the monthly allowances he has made to Amorette, beginning December 1 (a week after Missionary Ridge) and continuing until she came to stay at his house. He lists the butter and sausage he bought, the brandy he supplied, the watch repair, the charge for Seth's two attendants and their horses. He puts down eight small cash expenditures.

Then he comes to his largest item: "taking care of Deceased at my house & Boarding his wife 8 weeks from March 17th to May 19th 1864." He values his services as caretaker and landlord at seventy-two dollars. When he adds up all the items, large and small, he arrives at a total of $167.

Gathering up the papers, he goes out the door. We see him next in the courthouse, where his claim is notarized by an attorney, approved by temporary executor McAlvin (for of course Moses cannot pass judgment on the validity of his own claim) and allowed by the judge—all in the same afternoon.

A month later Moses makes the set-aside payment to Amorette. Though the state of Iowa intended it to provide support for the widow and child in the first year of their bereavement, Amorette does not receive it until that year is almost over.

Now Moses returns to his examination of the Austin claim, scanning receipted bills from Ohio, poring over Asaph Tuttle's credits for boots and shoes and his charges for buggy repair. Finally he concludes that the estate still owes Philander Austin six hundred dollars. Unfortunately only $450 remains in the account. Unwilling to deplete the small hoard entirely, Moses sends Austin two hundred dollars.

The scene changes to Mary Rich's house on Union Street.

Amorette walks in the door smiling. She is carrying a letter. The Second Auditor's Office in Washington has sent her a voucher for three hundred dollars, representing back pay that the Army owed Seth at the time he died. Since she was the one who applied for it, she believes it is rightfully hers and she intends to keep it.

Somehow Moses hears of her good fortune. He appears at Mary Rich's door and asks, then demands, that Amorette turn the three hundred dollars over to him. It is the rightful property of Seth's estate, he tells her. Unwillingly she gives him the money.

Two weeks later, apparently in retaliation, Mary Rich presents Moses with a claim of her own. The amount is staggering. Mary claims she has supported Amorette and Dell from the day in March 1860 when Amorette returned to her mother's home to have her baby until the day in March 1864 when she went to Moses' house to care for Seth. Now, over two years later, Mary is asking for compensation. The first item on her list of expenditures reads, "To board & clothing for wife and child four years at $4 per week—$832." Mary is also charging the estate ninety-nine dollars in interest on that amount. The other items are smaller: doctors' visits, bottles of Dandelion Bitters and Bark & Iron. All in all, Mary is billing the estate $961.

How Moses feels when he first reads her claim is written large on his face. He summarily dismisses it, telling Mary "that prior claims would have precedent in payment and there would nothing remain" to pay hers. He has been to a lawyer about the matter, he says.

Faced with his refusal, Mary retracts her claim and seems ready to give up, but then her son Angello comes to her defense. In an effort to extract compensation for his mother he begins to negotiate with Moses. Meanwhile Moses pays all of

Seth's back pay to Philander Austin. Now he can see his way clear to settling the dispute with the Riches.

On October 1, 1866, all the interested parties gather in Judge Rogers' courtroom to approve a compromise agreement allowing Mary $160. Mary and Angello sign it first, then Amorette, "Widow and Guardian of Dellizon Foote her son," and finally Moses as executor.

The end of Moses' task is almost in sight. That same week, back at his desk, he pays Austin the last installment, bringing the total amount Austin has received from Seth's estate to $774. Then he pays Mary all the money that is left in the account, which falls considerably short of the sum that was agreed upon.

A month later, and two years, five and a half months after his son's death, Moses appears in Judge Rogers' court for the last time. The judge approves the final settlement, stating that it conforms to the law in all respects. Moses has done his work efficiently. But the consequences are cruel for Seth's wife and child. Two-thirds of Seth's estate have gone to Philander Austin. Amorette, for whom Seth intended one-third, receives about one-sixth. Dell, who was promised two-thirds, receives nothing.

I do not blame Moses for his inability to carry out Seth's intentions; Seth was to blame for leaving his debts unpaid. Nevertheless the picture of Moses that emerges from this drama is, to my eyes, a picture of a hard-hearted man. His charge of seventy-two dollars for taking care of Seth and boarding Amorette offends me the most. It seems to me that caring for a dying child is not a service for which a parent should expect to be paid. Furthermore, I believe that Amorette earned her board while she was at Moses' house nursing her husband. But Moses seems to have shown little consideration for his son's wife, paying his own claim to the estate before he paid hers, even though the obligation to her took precedence.

Actually none of the three family members most concerned in the settlement came out of the long, wearying ordeal looking very good. Mary's claim, made in ignorance of the true condition of Seth's estate, was preposterous. And while Moses undeniably attended to his own needs first, Amorette probably had no legal right to Seth's back pay. Still, speaking now as her granddaughter, I wish she could have had it for herself and her son.

Chapter 15

Amorette's Battle

Her prolonged battle with the United States government had begun in February of 1865, when she filed an application for a widow's pension with the United States Pension Office in Washington. At that time she had not yet received the set-aside payment from Seth's estate, and she probably felt acutely the need of a reliable income. She had reason to expect some assistance from the government, since a law passed in 1862, granting pensions to disabled soldiers and sailors, also promised payments to the widows of servicemen who died in the line of duty.

The law, however, imposed several conditions upon applicants. As might be expected, the woman who claimed to be the serviceman's widow had to offer proof that she had been legally married to him. Furthermore, it would have to be established that he had been officially entered on the Army's muster roll, not only when he entered the service but also whenever he received a promotion. The evidence of muster-in to a higher rank was essential, for the husband's rank would determine the amount of his wife's pension. Finally, it would have to be proved that the disease or injury which caused the husband's death was contracted or sustained while he was "in the service of the United States and in the line of duty." Implicit in the third condition was the expectation that the soldier was in good standing with the service at the time of his death. Every one of these conditions would be difficult for Amorette to meet.

In the first place she had no legal proof that she and Seth were married. On the winter day, eight months after her husband's death, when she stood before a judge and watched him transcribe her answers onto a Widow's Declaration of Pension, she was forced to admit that there was "no public record" of her marriage, "only a private record." She could not even supply the name of the justice of the peace who married her and Seth. Evidently he had failed to make a permanent record, or the courthouse had burned down, or some such misfortune had eliminated any trace of license or ledger entry. I am not willing to entertain the idea that no marriage took place. Too many people, including her closest relatives, believed that it had. But their testimony was almost the only kind of evidence Amorette could offer.

Her brother Angello appeared in court with her. He had come from his home in Clermont to sign as witness to the Declaration, bringing along a friend, Ela P. Sanborn, who had agreed to do the same thing. Angello and Sanborn stated under oath that they were "well acquainted" with Amorette and Seth and knew "that they lived together as man and wife, and were so reputed."

Realto Price, whom Amorette had engaged as her attorney, was also present in court that day. He listened as Amorette carefully recited the facts the Declaration asked for. She reported Seth's enlistment in the 8th Kansas at Fort Leavenworth; the name of the regiment's commander, Colonel John A. Martin; the date and place of her husband's death. When the judge asked her for the cause of death, she reported "Ophthalmia and acute Rheumatism" because that was what Dr. Hamilton had said.

The completed Declaration was mailed to the United States Pension Office in Washington, and that office sent Amorette an acknowledgment of its receipt and assigned it a file number. Then the Commissioner of Pensions turned to the U.S. Adjutant General's Office for verification of Seth's military service. The reply, written at the end of March, was discouraging: "There is no evidence on file in this Office that Seth Foot was enrolled and mustered into service in Co. A, 8 Regiment of Kansas Volunteers." The Adjutant General's Office had, however, found one piece of evidence indicating that Seth had served. It was the report of his death, "Cause of Death not stated."

Someone—either an investigator in Washington or Amorette, assisted by Mr. Price—asked Kansas to provide evidence of a muster. Early in May the Adjutant General's Office of Kansas produced a handsome document, ornamented in gold with the state seal, certifying "on honor that according to the records of this office, Seth Foot, enlisted August 28th 1861. Was mustered-into U. S. Service Augt. 28th 1861 at Fort Leavenworth Kan. by Lieut. Affley, Mustering Officer. Was Promoted to Second Lieut. Co "A," 8th Regt. Kan. Vols, Infty, December 12th 1861. Died of Disease, May 14th 1864." The certificate was received in Washington, but it appears to have carried no weight with the officials who were investigating Amorette's claim.

Almost a year went by with no action. Perhaps the government's neglect of the needs of one soldier's widow should be excused, in light of the larger events taking place in 1865: the end of the war, the assassination of the president, the immense difficulty of constructing a just peace. But from Amorette's point of view, the official silence was unacceptable. In April of 1866 she and Mr. Price turned her claim over to an attorney's firm in Washington, hoping no doubt that someone on the spot could get the attention of whoever was in charge of Claim Number 84262.

Near the end of that year Angello provided a fuller statement about Amorette's marriage than he had made before. This time he said that Seth "recognized, lived and cohabited" with Amorette "as his lawful wedded wife" and that she was "reputed as such in the Community in which they resided." He ended with a mild disclaimer: "And further the deponent saith not." Naturally not. He was not present at their marriage; he had never seen a marriage certificate because evidently there wasn't one. He did the best he could for Amorette, ending with the assurance that he had no interest, "direct or indirect," in the prosecution of her claim. Wisely, perhaps, he did not mention the fact that the claimant was his sister.

No progress was made in 1867. In 1868, however, Congress passed a law that prompted Amorette to renew her claim, altering it somewhat in the process. The new law gave a widow, in addition to her pension, a small stipend for each child under the age of sixteen. Thinking that eight-year-old Dell was clearly covered, Amorette prepared a Widow's Application for an Increase of Pension. With Mr. Price's help she dictated a lengthy affidavit in which she tried to meet all the requirements of the 1862 and 1868 laws. She even felt it necessary to ask her mother for the Family Record, which she then sent to

Washington as proof of Dell's existence and legitimacy. It must have distressed Mary Rich to see it go.

Amorette's new affidavit was more detailed and more argumentative than her original Declaration. Though she was forced to repeat Dr. Hamilton's diagnosis of Seth's illness, she supplemented it with observations and opinions of her own, which Mr. Price couched in the language of the law. Seth, she said, died "of the disease known as Ophthalmia and Acute Rheumatism, caused by wound received at the battle of Mission Ridge in Tenn. and from bad exposure to the inclemencies of the weather and storms; which the said Seth Foot contracted while in said service and in his line of duty." She was no doctor; she did not know the term *pyemia,* though she had seen its effects at first hand. But everything she added to the doctor's diagnosis was true. Seth's disease was caused by the wound, and his exposure to the winter storms in East Tennessee probably contributed to the severity of the illness.

She reported that Seth was "on sick furlough at the time of his death," and asserted further that at the time he entered the service he was "a sound and healthy man." She had known his sound and healthy body, she had seen it consumed with the infection that eventually killed him. Now, she argued, "the disease of which he died he contracted in said Service and no other place."

With these documents Amorette and Mr. Price appeared to be building a persuasive case, but when the Pension Office asked her to provide a copy of Seth's furlough order, she did not see how she could comply, since she had sent it to the Second Auditor in Washington when she applied for Seth's back pay. Eventually, after going to "great trouble and expense," as she remarked, she procured a copy from the Auditor and sent it in. As evidence in her case, however, the order was not very convincing. Dated March 1, 1864, it announced a general furlough for the entire regiment; nowhere did it grant a sick furlough to Seth, as Amorette had claimed. On the contrary, it specified that Second Lieutenant Seth Foot, along with the other officers, would rejoin the regiment at Fort Leavenworth on April 4. His absence from his regiment for the rest of April and the first half of May would have to be considered as absence without leave unless mitigating circumstances could be found.

Nevertheless the Commissioner of Pensions, moved perhaps by Amorette's persistence, pressed on with her case. In April of 1870 he asked the Adjutant General's Office for evidence of Seth's muster and was told there was no evidence Seth had been mustered in as second lieutenant. By this time, apparently, Assistant Adjutant General Thomas Vincent was willing to accept the idea that he had been mustered into the service in August 1861, but the muster-in as second lieutenant was the important one, as far as Amorette's pension was concerned. Unfortunately, the certificate from Kansas, so positive in every other respect, did not include that phrase. Yet there might be a way around this difficulty, Vincent told the Commissioner. An "office muster" could be made, "if the facts in connection with the official records should so warrant."

At that point a serious search for the facts began, and by the end of October the Commissioner sent Vincent the evidence he had gathered. "Can you give this man a muster with help of the enclosed papers?" he asked.

Still not entirely convinced, Vincent asked the Paymaster General's Office from what date Seth was first paid as second lieutenant. "December 12, 1861," replied the Paymaster.

Vincent then turned to a member of his own staff: "How and when did vacancy occur for him as 2nd Lieut?" The question reached the desk of a man, working in Room 65, who signed himself JW. He had been given the task of searching the Kansas muster rolls for details of Seth's service record and copying out all relevant entries. "A vacancy in Co. "A" 8th Kans Vols.," JW wrote, "was caused by the transfer of 2nd Lieut. John Conover to Co. "I" same Regt.."

More notes went back and forth between offices until Vincent's secretary discovered that Seth had been, if not mustered, at least commissioned as second lieutenant. On December 21 Vincent concluded that he had sufficient evidence, and he dictated an order: "Make an office muster, mustering Seth Foot, into service as 2nd Lieutenant Co. "A" 8th Kansas Volunteers, to date from December 12, 1861, the date of his commission and from which paid."

Meanwhile the Pension Office had finally come to accept the legitimacy of Amorette's marriage to Seth. Angello's testimony had not entirely convinced them, but when someone back in Iowa, probably Amorette and Mr. Price, thought to submit a certified copy of Seth's will, the reference in it to "my beloved wife" tipped the scales in Amorette's favor.

Now two of the conditions for receiving a widow's pension have been met, but the third has not. It is the stipulation that the soldier's fatal disease or injury must have been contracted or sustained while he was "in the service of the United States and in the line of duty." Looking for pertinent information, JW continues to study the Company A muster rolls, filling sheets of ruled paper with his notes. As he scrutinizes the roll for the spring of 1864, he comes upon a startling entry: The returns for April and May 1864 report Seth "Absent without leave since Apl. 4. '64." JW copies this disquieting statement into his notes and follows it with the report, taken from the June return, that Seth had died "of disease." Then he feels compelled to add, "Charge of absence without leave not removed." What this discovery means is that for the past seven years Seth has been carried on the Army records as a deserter, of whom there were so many in that punishing last year of the war.

JW shows his page of notes to E. E. House, his superior, who notifies Thomas Vincent of this new problem. A search is begun for evidence that would mitigate the AWOL charge, and Dr. Hamilton's two medical certificates, dated April 2 and 22, 1864, are found. Seeing them, Vincent issues another order: "Remove the charge of absence without leave, entered against 2d Lieut. Seth Foot, on the April Return of Co. A, 8th Kansas Vols. for 1864, medical certificates having been filed in this office which properly account for his absence from command from April 4, 1864, to May 14, 1864, the date of that officer's death." In Vincent's measured recital of facts, ending in the one irreversible fact, I hear an undertone of pity and scorn—pity for the dead officer, scorn for the incompetence that had sullied his record.

Back at his desk in Room 65, JW comes upon the entry that covers the month of Missionary Ridge: "Roll for Nov. & Dec. '63 - 2" Lt. absent on D.S. as Ord. Off. 1" Brig. 3d. Div. 4' A.C.   at Battle of Mission Ridge Nov. 25. '63" [Roll for November and December 1863 - 2nd Lieutenant absent on Detached Service as Ordnance Officer, 1st Brigade, 3rd Division, 4th Army Corps   at Battle of Missionary Ridge November 25, 1863]. Without giving that crucial time period any special thought, JW goes on copying subsequent entries, filling that page of notes and beginning a new one. Nearly two weeks later, as he is reexamining the rolls, he sees something he had not noticed before: on the entry for November and December 1863 a word appears to be missing at the end of a line, after "4' A.C.." Turning to his notes, he inserts a carat between "4' A.C." and "at Battle of Mission Ridge." Above the carat he draws an X, which he shades in red ink. At the bottom of his second page he draws a second X, shades it red and writes, also in red, "There seems to have been another word after 4th A.C. at the end of the line, but the edge of the roll is torn away and there is nothing in Returns or subsequent rolls to show what it was." He passes the question on to Mr. House.

House guesses what the missing word may be and writes a colleague, "Please state if he was wounded at Mission Ridge. I think the word 'wounded' may have been omitted in this report by mistake."

House had guessed right, of course. For the preceding seven years, the Army had had in its possession, somewhere, two written records of Seth's wounding. The first was the List of Casualties of the 8th Kansas for November 23-25, 1863, the days when Orchard Knob, Lookout Mountain and Missionary Ridge were retaken by the Union. On it Colonel Martin had reported that Lieutenant Seth Foot was "severely" wounded in the "thigh." The other written record was General Willich's report of the battle of Missionary Ridge, written on December 31, 1863. Willich had mentioned Seth's wounding parenthetically, almost offhandedly: " . . . ordnance officer, Lieutenant Foot (who on this occasion was wounded by a shell) . . . " But he did say it, and he submitted his report to the man (not Vincent) who was acting as Assistant Adjutant General at the time. Evidently Vincent never saw either Willich's or Martin's reports.

The question of a battle wound was essential to Amorette's case, as all the investigators knew. Accordingly, a search was made of the muster roll returns from the First Brigade (General Willich's brigade). But the answer came back, "No evidence of having been wounded."

With that avenue closed, House evidently felt he had done

all he could, for he wrote the Commissioner of Pensions, "It appears from rolls that on the 25th Nov. 63, he was at battle of Mission Ridge, on detached service as Ordnance Officer of the Brigade, but whether wounded, not stated." House signed his initials at the bottom of his memo and then decided to add one more sentence: "Med. Cert. dated April 22, 1864, upon which his leave was extended, states that he was 'suffering from Rheumatism and extreme debility.'" Under this message, which with its vague diagnosis further weakened Amorette's claim, Vincent added his initials.

There her case stood in the spring of 1871, and there it might have stood indefinitely, for nothing in either of Dr. Hamilton's 1864 certificates indicated that Seth's illness had been contracted in the line of duty. But between the middle of April and the first week in June new evidence came in, convincing evidence that Seth was wounded at the battle of Missionary Ridge and that his death was the consequence of the wound.

An eyewitness account of the wounding of Second Lieutenant Foot came to the Pension Office from Sol Washer, former adjutant of the Eighth Kansas. During Seth's illness Washer had sent Seth a letter containing useful advice and friendly encouragement. Now he offered unequivocal evidence of Seth's wounding. His affidavit, as summarized by the Pension Office, testifies that Seth "was wounded at Battle of Mission Ridge Nov. 24, 1863 in right hip by, as affiant believes a piece of rock, scattered by a cannon ball—wound received while in action and line of duty—was present at & during said Battle and have personal knowledge of facts."

But though Washer's testimony proved that Seth was wounded, it did not explain why a wound received in November would cause a man to die the following May. That explanation came from Dr. Hamilton. In a new affidavit, directed to the Pension Office, he testified that "he was called up to attend said soldier March 26 1864 and continued to attend him until May 9 1864 about time of his death. He died of Pyemia caused by the absorption of Pus into the circulation—Pus formed near the bone and failing to reach the surface passed into the circulation—came from a deep seated abscess caused by contusion of right hip. Died of Pyemia so caused. Died May 14, 1864."

Here, finally, was a convincing chain of testimony. Washer saw the rock hit Seth's right hip. Dr. Hamilton saw the abscess in the right hip, recognized pyemia and seven years later said so.

Who had the idea to call upon Sol Washer and Horace Hamilton for testimony? The record does not say. But it was Amorette who had kept Washer's letter in her possession after Seth died. Probably she had seen the letter when it arrived; she may have read it to Seth as he lay in bed. Now, when the testimony of someone who had known Seth was so badly needed, she may have thought of Washer. Was she also the one who asked Dr. Hamilton to submit a new medical report? She knew that Seth had been wounded. Did she ask the doctor to confirm for the investigators what she had known all along? I think that Amorette, aided by Realto Price, was the most likely person to have broken through the logjam in Washington.

Six and a quarter years after she filed her original Widow's Declaration, Amorette was granted a pension of fifteen dollars a month for life, unless she remarried. The stipend for Dell was denied, which was disappointing, but at least the payment of the pension was to be calculated from the day after Seth died. The retroactive payments would come to $1,260, a sum that would benefit both Amorette and Dell.

Amorette's efforts in the six-year campaign accomplished

two more things. They caused the unjust charge of absence without leave to be removed from Seth's army record, though she probably didn't know it was there and wasn't told when it was removed. Moreover, the investigation that she provoked prompted Dr. Hamilton to make an accurate report of the cause of Seth's death—a small advance in the annals of medical science. I think my father would have been interested to know Dr. Hamilton's final diagnosis, but I don't think he did, or he would have told my mother and she would have told me.

Others beside Amorette deserve credit. Realto Price for staying with the case for so long; Thomas Vincent for his forthright, humane dismissal of the AWOL charge. JW for noticing that a word was missing on the muster roll; E. E. House for guessing what it was. Dr. Hamilton for allowing his informed medical judgment to prevail at last. And Sol Washer—of course Sol Washer. He said he saw a piece of rock wound Seth. Maybe it was a piece of rock or maybe, as General Willich believed, it was a shell. It did not much matter. Washer's testimony, along with that of Dr. Hamilton, enabled Amorette to have a lifetime income, small but dependable. Clearly she had earned it.

Dell and Amorette, about 1871

Chapter 16

The Proposal

The summer after Amorette was awarded her pension began with a baseball game and ended with a circus. In early June the Fayette Nine played the West Union Nine. It was a home game for Fayette, so that Dell might easily have been in the crowd of spectators. I know he played baseball himself when he was a little older. Many years later he told me so, showing me the finger, permanently bent, that he had injured in the sport. He was probably too young—eleven that summer—to have played with the Fayette team, but he could have been at the game, watching, enjoying, learning.

There is a picture of Dell, taken at about that age, in Mary Rich's photograph album. It shows a boy of slender build and sober expression. His lower lip protrudes slightly beyond the upper one, and he has a receding chin—the kind of chin novelists used to call weak, suggesting a weakness in character. But there is no reason why a boy's character should be either determined or reflected in the bone structure of his face. When I compare this picture with the one Seth sent home from Leavenworth, I see that Dell has his father's mouth and chin. Seth wore a goatee, which made his chin appear more assertive than it actually was. His son would grow a beard, as soon as he could, to accomplish the same thing.

The circus, with its train of wagons, clowns, musicians and animals, rolled into Fayette on the first Sunday evening of September. As the work crew was setting up the tents, a crowd of citizens gathered to voice their objections. This was the sabbath, they said. The Ten Commandments taught that no work should be done on the Lord's day. They cited the town's Sunday laws.

The town constable proposed a compromise, explaining to the circus manager that the people of Fayette "desired to see as little work as possible going on." The manager, who for his part desired not to offend potential customers, promised that his crew would do no work that night except what was absolutely necessary for the comfort of the animals. Mollified, the citizens returned to their homes. On Monday, the last day of summer vacation for the schoolchildren, the circus train paraded through town and into the big tent, and the show was on. It would have been a glorious day for a boy, if his mother's religious scruples did not prevent him from attending. The next morning the children—with Dell, no doubt, among them—trudged, ran, skipped and dawdled along the roads from home to school. By the time they were all in their seats, summer was over.

Autumn, however, brought its own pleasures and satisfactions, one of which was the apple harvest. The year before, Amorette's brother Angello, a nurseryman, had sent the editor of the *West Union Gazette* some apples, along with a letter describing them as "crisp, tender, juicy, with brisk, aromatic flavor." He told the editor the apples were "full medium size, when you consider that apples are unusually small this year." Let us hope that they were larger in the next year's harvest, and that Angello shared his with his mother and sister and nephew. Dell always liked apples, I know that. Once, when he was quite old, he showed me how he could break an apple in two with his bare hands. Perhaps he had learned the technique from Uncle Angello.

When winter came to Fayette, it was enjoyable for a while. The cold spell in early February packed the snow hard and smooth on the roads, so that sleighing was good. But winter was slow to relax its grip that year. The maple sugar season could not begin as long as there was frost on the ground, and in early April the frost was still there. On the morning of Dell's twelfth birthday a fierce storm of snow and wind blew in. It lasted all that day and the next, and the snow stayed on the ground for nearly a week. Then, suddenly, the air softened and spring arrived. "Soap is boiling," the *Gazette* reported, "and barefooted children are raking up the chips." The chips the children were raking must have been wood chips, used in the soapmaking process. I wonder if Dell was one of those barefooted children or if he would now have considered it beneath his dignity to be classed among them.

The boy in the photograph does seem to possess an air of dignity. Perhaps it comes from the way the photographer has posed him: one hand in his lap, the other arm bent, resting on a table. I caution myself not to draw too many inferences from what may have been merely the custom of the times. But there is also something in his eyes, his expression; this boy seems to have seen a good deal and kept it within himself.

I look at another photograph, one of Amorette, taken at

roughly the same time, and give myself the same caution. I see a certain primness about the woman in the picture but much of that may be attributed to the fashions of the early 1870s. Her dark hair is drawn back rather tightly behind her ears, though the curliness of it and its glossy sheen soften the severity of the style. She is snugly clothed. A high lace collar hides her neck almost completely. Her jacket, pulled tightly across her bosom, buttons over her figured blouse. Her lips are closed, and already I see a slight downturn at the corners. Her skin is clear and smooth, but with a slight heaviness along the jaw. Her chin, unlike her husband's and her son's, is long, with a slight forward thrust. My hypothetical novelist might conclude from this last detail that Amorette was a determined woman, but I will not admit a physical feature into evidence. And I happen to know that on one important matter she was indecisive.

The important matter was a proposal of marriage. My mother told me the story when I was young. She did not know what the suitor's name was, how Amorette came to know him, whether she cared for him. Mother's point—and of course this had been my father's story originally—was that after Amorette received the proposal she asked him (her son Dell, that is) whether she should accept it. Dell told her no. And so Amorette declined the man's proposal.

Mother laughed as she told me the story. She did not believe in asking children to make important decisions. "Just imagine," I can hear her say, "asking your child whether you should marry a man!"

At the time I shared Mother's amusement. The story suggested a weakness in Amorette's character, a lack of faith in her own judgment, a habit perhaps of excessive deference to the wishes of her only child. But now I am not so critical. I remember a time when I was faced with a choice similar to Amorette's.

My son Robbie was eleven years old when Cap asked me to marry him. I never considered asking Robbie whether I should accept the proposal. I'm sure that if I had, he would have said no. But I married Cap, and the antagonism between him and Robbie, visible even before our marriage, increased. Finally, despairing of making peace between my husband and my son, I sent Robbie away to school and then grieved over his departure. That was not the kind of mother I wanted to be.

Fear of conflict between her son and a stepfather may have been what moved Amorette to ask Dell's permission and to give up the idea when he said no. Of course there could have been some other reason. She may have been reluctant to lose her hard-won pension, but fifteen dollars a month was not very much money, even then. She might not have cared for the man, but in that case why would she have consulted Dell? It looks very much as though the original explanation, handed down from Daddy to Mother to me, was the correct one: Amorette put her son's wishes and needs before her own. I think that is the kind of mother Amorette was.

Amorette never remarried. I have been married four times; my first three marriages ended in divorce. This is the greatest difference between my grandmother and myself, the thing that would make it hardest for the two of us to understand each other, if we were on earth at the same time. I am not proud of my personal history, but I would not exchange it for Amorette's if I could. Yet I want to sympathize, if I cannot fully empathize, with the choices she made. That is one reason I am writing this book.

Chapter 17
Keeping House

When the 1880 census was taken, Amorette's occupation was listed as "Keeping house." It was the first time she had officially been given this designation. In 1870 it had gone to Mary Rich, while Amorette was listed only as one of the other occupants of the household. But Mary died in 1877; after that, naturally, the title devolved upon Amorette. What it meant was that she was the person who was responsible for the care and management of the house—in other words, the one who was expected to practice what Catharine Beecher had called "domestic economy."

Moreover, Amorette had found a way to make her occupation pay: she was taking in college students as boarders. At the time of the census she had two, a music student named Carrie Barden and a science student named Albert Bernau. If Amorette charged the rate the university suggested, she would have added as much as twenty-five dollars to her monthly pension income.

Dell was listed on the census form as both a son and a student; he, like the other two, was enrolled at Upper Iowa University. He entered at the age of twenty, as a junior. (It sounds as if he was given advanced standing for some reason.) The college catalogue lists him several times—first under the Collegiate Department, then under Commercial, Music and Art, but never under the name of Dellizon. He is always shown as Delano Angelo Foote. His evident dissatisfaction with the name his mother had given him must have hurt her feelings.

Dell too was bringing in some money. He paid his college expenses, as he once told me, by firing people's furnaces and playing the organ. He did not tell me what organ he played, but I think it is likely to have been the one in the newly built Methodist Church—the one the Young People's Society bought after the adults were unable, or unwilling, to lay out any more money. An old photograph of the interior of the church shows a reed organ, equipped with a row of stops above the keyboard and with foot pedals to operate the bellows. I picture my father (forty or more years younger than I ever knew him) sitting before the instrument, his hands moving over the keys, his feet vigorously pumping the air that kept the hymns alive.

If we can judge by the photographs in Dell's album, almost all of his fellow students were white. But I know of one African-American student who was in attendance at the same time he was. Susan Collins probably did not attend the same classes as he did, since she was enrolled in the Preparatory Department and he in the Collegiate. Nonetheless I think he would have been aware of her presence because the student population was small—281 students that year—so that a "colored person" would have stood out.

Born in 1851, Susan Angelina Collins was twenty-eight when she enrolled at Upper Iowa University. A photograph of her, kept in the university's archives, shows that in dress she resembles the young women in Dell's album: the round white collar, the ribbon bow at the neck. Like many of them, she has parted her hair in the middle and drawn it back behind her ears. But she looks different, and not just in the matter of skin color. There is a brightness in her face, almost a smile, an expression not seen in the others, who appear to have assumed the solemnity incumbent on college women. This daughter of former slaves does not really look older than the white students, though she probably was, but she looks readier, somehow, to go forward into the future with a cheerful countenance.

Susan's father, a slave who had escaped into freedom when his master took him to Illinois, had served in a colored unit in the Civil War. About six months after the war ended, he brought his wife and children to Fayette County, settling near the "Negro" enclave of Frog Hollow. Susan attended public schools and after leaving school worked as a maid in the homes of several Fayette families. But I am fairly sure she never worked for Amorette because I do not think Amorette could have afforded household help. Coincidentally, Susan and Amorette shared the same birthday, July 3. The lives of these two women, who probably did not know each other while they lived in Fayette, would follow a parallel course for a time after they left it. Then their paths would diverge widely, only to intersect in California on a day in 1901.

Susan Collins left Upper Iowa University at the end of Dell's junior year. If she had stayed another year, he might well have been her English teacher. For in his senior year, 1881-1882, he served as instructor in English, the only person teaching English that year. It was customary at that time for a few selected seniors to do some of the teaching. He received a bachelor's degree that spring, but whether it was in English or Commercial Studies is not clear to me.

Armed with the new degree, Dell decides to try his luck in Dakota Territory, a great rectangle of prairie, plain and mountain whose southeastern corner touches the northwestern corner of Iowa. Amorette will go with him, of course. But Fayette has been her home for over twenty years. We can assume that she is comfortable in the house; she has her friends in the church; her brother Angello lives nearby, along with his wife Isabella and their daughter Eleanor, who is twenty this year. It will be hard for Amorette to leave.

Now forty-three, she again faces a situation she had encountered when she was Eleanor's age: the man she loves and lives with tells her he is going west to seek his fortune. There has even been another gold rush, though Dell has not evinced any interest in joining it, and the excitement over the Black Hills mines, so strong in the previous decade, has died down by this time. Now land, free land, is the magnet drawing adventurous men and women to Dakota Territory. But Amorette does not necessarily partake of that adventurous spirit. The prospect of setting up housekeeping in a place that is not even a state, an uncivilized place, home to whole tribes of Indians, is likely to intimidate a middle-aged white woman. And what do she and Dell want with 160 acres of unturned land when neither of them has ever done any farming?

Chapter 18

D. A. Foote, Deputy

The Footes and Riches, in their westward migra- tions, had never been trail-blazers; they did what a great many other people were doing. True to form, Dell and Amorette went to Dakota when a land boom was in progress. In the 1880s there were more land entries for Dakota than for any other territory or state. The homestead policy, first put in place during the Civil War, entitled any person—man or woman—to make a claim for 160 acres of public land. To qualify for ownership, the claimant was required to reside on the land for five years and make improvements on it. Dell and Amorette probably did enter a claim, just as my mother had told me. So did another Fayette resident; in November of the same year, 1882, Susan Collins came to Dakota. Both she and Dell found it necessary to work at something besides homesteading. Susan opened a laundry. Dell, calling upon the commercial skills he had learned in college, got a clerical job in the new town of Huron. At some time during his stay—probably in the last year—he was appointed Deputy Clerk of the District Court of Beadle County (Huron's county).

Fortunately for me as chronicler of my father's early career, his name found its way into two public records. The first was a scientific report. On August 28, 1884, a tornado approached Huron, threatening to demolish it. The progress of the twister was observed by (among many others) Sergeant Samuel Glenn of the Signal Corps, who set down his observations as promptly and vividly as he could. The tornado first appeared as a white funnel in the northwestern sky. "With the swiftness of thought," wrote the poetical sergeant, "it spun out toward the ground a white, ribbon-like line."

The fearful white tornado seemed to be heading directly toward Huron. But after making a venomous hissing sound, it changed direction, so that instead of cutting a swath through the town it touched down at a farm three-quarters of a mile away, where it destroyed the farmer's barn but not his house. The tornado's change of direction allowed the townspeople, or at least those who had not taken refuge in cellars, the privilege of watching the natural phenomenon from a position of safety. A good many of them did so. Dell was one of them.

Soon after the storm was over, Sergeant Glenn made a series of sketches to show the stages of the white tornado's progress, appending them to the report he was writing for his superiors in Washington. Before he submitted it he solicited the signatures of witnesses to certify that his sketches accurately depicted the phenomenon. First to sign were the editors of both newspapers and a member of the House of Representatives. Then came the mayor, two bank presidents and the president of the Central Dakota Medical Society. Next were the Register of Deeds, the Deputy Register of Deeds and the Chief Clerk of the U.S. Land Office. After that distinguished beginning the signers were a more mixed group: a man who wrote his occupation as simply "lumber," two attorneys, an engineer, a "Police Justice," the superintendent of the Western Union Telegraph Company and a bank cashier. The Reverend A. P. Graves, D. D., the only clergyman among the forty-eight witnesses, signed next. Below his name Dell wrote "D. A. Foote, Clerk."

The other public record consists of forty-six declarations of intent to become a citizen of the United States. Dell's signature as deputy clerk appears on each one. This collection gives us a picture of Dell as he worked at his desk in the county courthouse. It also allows us to see into the great wave of immigration that swept over the nation in the eighteen-eighties and to discern the outlines of some of the men and women who were swept up in it.

The Dakota land boom peaked in 1884, but in early 1885 immigrants were still entering the territory in large numbers, many of them coming in from other countries. Between late January and early June, 1885, Dell interviewed forty-six immigrants and penned their declarations of intent. He signed his name to show that he was the one who had filled in the blanks, though the official signature was always that of the County Clerk. To the left of the Clerk's name Dell would write his own. At first he stood on his dignity, signing "Per D. A. Foote Deputy." But later he dispensed with the Latin word and abbreviated his title so that his signature read "By D. A. Foote Dep'y."

Of the forty-six declarers, thirty-seven were male and nine were female. Five were unable to write their names, signing with an X, which was witnessed by Dell and one other person. Four of the five illiterate signers were women. Fifteen of the immigrants were born in Germany and sixteen in Great Britain, Ireland or the Dominion of Canada. Seven were born in Norway, three in Sweden and two in Denmark. Two were born in Belgium and one in France.

Dell appears to have written down exactly what each person told him, even when, as sometimes happened, it did not make historical or geographical sense. The declarers, as a proof of seriousness, had to certify their intention to "renounce forever" all allegiance to any foreign ruler or state and specify the person or entity concerned. All but two of them promised to renounce their allegiance to monarchs, and some named the monarch: Oscar II of Norway and Sweden. . . Christian II King of Denmark. . . Wilhelm Emperor of Germany. . . The Queen Victoria Queen of Great Britain and Ireland. The other two specified an elected official, the President of France. Nothing surprising about that choice, except that one of those who said he would renounce allegiance to "the President and Republic of France"—a man named Clement de Somer—also reported that he was born in Canada. His answer is puzzling, since France had lost its possessions in Canada over a hundred years earlier. But Dell wrote down Monsieur De Somer's response with no comment.

Most of these immigrants had entered the United States at New York, as would be expected. The next largest group, five Canadians, gave Huron as their port of entry. All but one of the others came in from Detroit, Boston or Baltimore. The one exception, an Englishman named Henry Richards, said he entered the United States "at the port of Quebec." Dell faithfully wrote Richards' words into the record. It seems he did not always have his mind on his work.

With the warming of the Dakota weather the number of declarers increased. Where there had been only two in the entire month of January, there were seven in the first three days of June. The declaration of a twenty-one-year-old man from Germany, dated June 3, was the last one signed by D. A. Foote, Deputy. The sudden stop, as the tide of immigration was steadily rising, suggests that early June was the time Dell stopped working for the District Court.

He and his mother were leaving the territory, going back to the States. Not only that, they were going to Chicago, the metropolis of the middle west. It was a bold move for these two, who up until this time had never lived anywhere but in small towns. Dell had decided to enroll in a homeopathic college of medicine, another bold move, for the homeopathic approach to medicine was preached and practiced by a small minority of doctors and was ridiculed by many of the "regular" M. D.'s. Looking at his college record, we might have expected him to choose a career in business or in the teaching of English. His work for the District Court showed him what he could do as a civil servant. Yet he decided to study homeopathic medicine in Chicago. We are led to wonder what, or better yet who, could have put the idea in his head. I think it might have been Dr. Primm.

J. W. Primm, M. D., with offices on Dakota Avenue in Huron, advertised himself as a "Homoeopathist." He gave as a reference a Professor T. C. Duncan, M. D., whose address was on South Clark Street in Chicago. The first homeopathic medical school in Chicago, the Hahnemann Medical College, was located on South Clark Street. Very likely Dr. Primm himself had graduated from that institution. It sounds to me as if Dr. Primm encouraged Dell to follow his own example.

Homeopathy was conceived in Germany by Samuel Christian Hahnemann as a reaction against the ignorant, often cruel treatments administered by the medical profession, particularly the reliance on calomel (a preparation derived from mercury) and the widespread practice of bleeding a patient in order to cure him. The homeopathic way of thinking came to the United States early in the nineteenth century. In the 1870s and 1880s it was still seen as the "new school" of medical practice—arousing suspicion in the older doctors but intriguing some young men, and perhaps a few young women, who were considering medicine as a career. Evidently Dell was one of those young men.

Relinquishing their ambition (never very serious, I think) to be homesteaders, Dell and Amorette entered into the next phase of their life. I am still using the singular—it is still their life, not their lives. Eventually they will go in different directions, but not yet.

At this point in my story, while Amorette is waiting in the wings, so to speak, before she steps out upon a larger stage, I want to make an admission about myself. For several years after the burglary I remained angry and resentful toward my stepson Jon for stealing her watch, if that is what he did. I especially hated the thought that whatever money the watch had brought him should have been spent on heroin. It was as if not only the watch but also my grandmother herself—my pious, sheltered grandmother—had been debased, dragged down into a world that she could not possibly have recognized.

I am glad to say, however, that I do not feel that way any more. I remember the day I realized the anger was gone. It was a remarkably clear day, I recall. I had just walked out the front door of the Graduate Theological Union Library, which sits at the top of what Berkeley people call "Holy Hill." I was carrying an armload of books on nineteenth-century women in the Methodist Church. I had found Amorette's picture in one of them and I was elated. I stood still for a moment, looking westward. The rare sight of the bay, the bridge, the islands, the mountains—all standing out with the utmost clarity—sharpened my inner sight as well, and I realized that if it hadn't been for the burglary I wouldn't have been impelled to learn as much about my grandmother's life as I now knew. That knowledge, I saw, was more valuable to me than the watch.

PART FIVE

"All the Word 'Mother' Implies"

Resident Faculty, Chicago Training School, 1890
Courtesy of the Graduate Theological Union Library, Berkeley, California.
Amorette is second from the right in the second row.

Dell in Chicago, about 1888

Chapter 19

Finding Her Vocation

In the fall of 1885 Dell enrolled in a homeopathic medical school, but not the one on South Clark Street. Instead he chose the Chicago Homoeopathic Medical College, founded nine years earlier by a group of men who had broken away from Hahnemann. They seceded, they said, in order to provide a more rigorous medical education, still keeping to the homeopathic tradition. Their action sounds like a response to the criticism, often voiced by "regular" medical men, that homeopaths were ill-educated quacks.

The two fundamentals of homeopathy were the law of similarities and the law of infinitesimals. The law of similarities taught that a disease could be cured by drugs which in a well person would produce the symptoms of the disease. The law of infinitesimals called for exceedingly small doses of the prescribed drug—the smaller the better, according to Hahnemann's most faithful disciples. The two beliefs, while seeming absurd to most members of the medical establishment, had the virtue of at least doing no harm to the sick person. That feature alone recommended itself to many patients. Gradually the two schools of thought grew closer together: the regular doctors abandoned the bleeding and the harsh medicines they had formerly employed, while homeopaths began prescribing those of the regular doctors' drugs that appeared to be both effective and safe. By the time Dell entered medical school, homeopaths had gained acceptance from many of the regular M.D.'s.

I doubt if Amorette understood the philosophical foundations of the profession her son had chosen; nevertheless she came to know the effect his choice would have on her life: it brought her to the city in the first place, and it would leave her there, alone. After completing his two-year course of studies, Dell spent the following year serving as an intern at a homeopathic hospital in Chicago and working as a prison doctor at Joliet State Penitentiary, about thirty-five miles away. At the same time he was looking around for a place where he could set up a permanent practice. He found one in Omaha. The medium-sized city, thirty-four years old in 1888, lay along the banks of the Missouri, on the eastern edge of Nebraska. To get there from Chicago Dell traveled all the way across Illinois and Iowa. This time Amorette did not go with him.

With Dell gone, Amorette moved into a new home and undertook a new occupation. At the opening of the school year 1888-1889 she began work as Matron of the Chicago Training School, established three years earlier by Lucy Rider Meyer and her husband Josiah Shelly Meyer. Its purpose was to prepare women to be Christian missionaries. The school, which was connected with but not supported by the Methodist Episcopal Church, was located at the corner of Ohio Street and Dearborn Avenue, five blocks north of the Chicago River and seven blocks west of Lake Michigan. State Street, a principal thoroughfare, and Clark Street, where a Methodist church was located, were each only a block away. "The location was chosen," Rider Meyer wrote, "because it was in a good residence locality, and yet within a few minutes' walk of the street-car lines to all parts of the city."

In searching for a "good residence locality," the Meyers were thinking, no doubt, about the safety of their students, most of them single women who had come to the city from small towns and rural areas. One of them was Susan Collins, who left Dakota in November of 1886 to enroll in the new school. She had learned of it one day while she was working in her laundry. A customer brought in a bundle of dirty clothes wrapped in a newspaper. Unwrapping its pages, Susan saw an advertisement that urged women to come to the Chicago Training School and prepare themselves for Christian service. She answered the summons and entered the program. She was not there when Amorette arrived, however. In the spring of 1887 she had gone to Africa as a missionary.

By the time Amorette began her work, the school had added a new, more specialized kind of training program. It began almost by accident. In the summer of 1887 a few members of the first graduating class stayed on after graduation to continue the regimen of home visits that had been part of their curriculum while they were students. Their work that summer, bringing the gospel of Christ, as well as practical assistance of many sorts, to poor people in Chicago's slums, led to the formation of the first active deaconess program in the Methodist Church. Early in the life of the organization it was decided that deaconesses should wear a distinctive garb, "for protection," as they went about the streets of Chicago doing their missionary work. It would consist of a plain black dress and a small black bonnet, worn high on the head and tied under the chin with white ties. Besides identifying the wearers as charitable workers, the uniformity of clothing expressed "sisterly equality" and "sisterly union," Rider Meyer said.

The Methodist deaconess work that began in the autumn of 1887 responded to a felt need for religious/charitable outreach in America's cities. In Chicago, in 1890, immigrants would constitute over forty per cent of the population. They crowded by necessity into the cheapest, dirtiest, most ramshackle sections of the city. Many were Catholics, others were Jews, still others professed no religion at all. In these newcomers the Protestant churches saw a dual mission: to relieve the suffering of the immigrant poor and to convert them to Protestant Christianity.

Amorette did not enter the deaconess training program when she first arrived at the school. Her job as matron was to keep house on a scale far larger than any housekeeping she had done before. Her qualifications for the position were modest. At least once in her life she had run a boarding house, though it was a small operation compared to this one. The fact that she was a longtime member of the Methodist Church would also have recommended her, as would her willingness to work without pay.

But beyond these material considerations, I believe there was something about Amorette that evoked the image conveyed by the word *matron*. Its root derivation is the Latin *mater*, mother. In ancient Roman times *mater* had already given rise to cognates, of which one was *matrona*, meaning "a married woman, a woman of character." *Matrona loris* meant "the lady of the house." In nineteenth-century English, *matron* often denoted a woman who was in charge of the domestic affairs of an institution, and this is the meaning that applies directly to Amorette's new position. The original Latin root, however, should not be ignored. Lucy Rider Meyer, very likely the person who engaged Amorette, cherished the concept of the mother, as did other leaders of the Methodist Church. Meyer's own mother had filled the role for the first class of Training School students. "She is the real mother in our home," Rider Meyer wrote. After the first year the older woman was no long able to serve in that capacity, but the need for a mother-figure continued to be felt. I believe Amorette was chosen not only because of her boarding house experience and her Methodist loyalties but also because she seemed to possess the maternal qualities that the job of matron demanded.

A stated purpose of the Training School was "to furnish a true home under Christian and missionary influences for our students, and to give practical training in domestic work and management, under the care of a competent matron." It was Amorette's duty to make that true home operate smoothly and to give that practical training. Rider Meyer clearly considered her to be competent, calling her "our efficient matron" or "our good matron" whenever she had occasion to mention Amorette's name in print.

The building that housed the school, newly built through the fund-raising efforts of Mr. Meyer, was four stories high, though only twenty feet wide. The entrance, framed by a stone arch, led into the first floor, where the reception room, the office, the library and the recitation rooms were located. Sleeping rooms for the students and the resident faculty occupied the other three floors. On one of those upper floors Mr. and Mrs. Meyer had their apartment.

These facts about the building's interior come from an article by the Reverend J. C. Jackson, who toured it from bottom to top shortly after it opened. The basement housed not only the dining room, the kitchen and washroom but also a free dispensary, with separate entrance, to treat the illnesses of "the worthy poor," Jackson said. It is hard to know whether the distinction between the worthy and unworthy poor was made by the leaders of the Training School or only by the Reverend Jackson himself. I do know that the deaconess organization, which was formed about five months after his article was published, ministered to the "unworthy"—drunkards and outcast women, for example—as well as to the more virtuous.

The basement was the place where the most intensive housework was done. The kitchen must have been busy much of the time, as food for thirty or forty people was prepared and served. It was plain food, Jackson said: "Pastry there is almost none; meat comes once a day." Amorette was probably the one who planned the menus for those plain meals, a task made more difficult near the end of her second year when the Training School came close to running out of money. All the residents—students and faculty alike— ate together and together observed the religious rituals—the prayers, the hymns, the Bible readings— designed to keep alive in each heart a sense of their shared mission. The domestic chores, however, remained to be done. After each meal a small crew of students (each wearing the kitchen apron she had brought from home) carried the dishes back to the kitchen and washed and dried them, by hand of course.

At intervals, perhaps once a week, the tablecloths, napkins, dishcloths and dishtowels would find their way to the wash-room, where they, along with the sheets, pillowcases and whatever other laundry had accumulated, would be scrubbed, boiled, rinsed and hung to dry. Then there was the cleaning—the dusting, sweeping, mopping and scrubbing of the common rooms on the first floor and the sleeping rooms and bathrooms above. Taken altogether, the domestic labor must have consumed a good deal of the students' time and energy, but at least they were prepared for it before they decided to enroll. The catalogue made it clear that each student was expected to give "cheerfully" about an hour a day to "the general housework of the home."

The Training School did not employ any paid domestic workers. The chief reason was that it did not have the money to do so. But there was another reason, the Reverend Jackson said, for requiring these future missionaries to do the housework. It was because "the mode of life to which they must become inured is necessarily one of self-denial." There was a spiritual component, then, in the work requirement, and it was Amorette's job to reconcile the spiritual and the practical demands. She had to see that the housework was done, done properly and, whenever possible, cheerfully. The matron at the Chicago Training School, like the nineteenth-century mother in her own home, was expected to offer moral guidance and set an example of Christian living for the "daughters" in her charge. I think Amorette would have accepted this obligation willingly.

One day, pulling Lucy Rider Meyer's history of deaconesses down from a library shelf, I came across a picture of my grandmother. There were six other people, one man and five women, in the photograph. The caption identified them as the "Resident Faculty." Though no names were shown, I recognized my grandmother right away, for I already knew her face. Amorette is standing next to the end of the back row on the righthand side. She is looking soberly off toward something outside the camera's range, as nearly all the other people are doing. The man leans back slightly, and he has crossed his arms over his chest—a commanding posture. He is Josiah Shelly Meyer, superintendent of the school. Mrs. Meyer, its principal, stands directly before her husband. She is a thin-faced woman with a thoughtful expression.

Amorette appears older than any of the others. The line from her nose to the corner of her mouth is deeper, and something about the set of her head and the slope of her shoulders says that she has felt the weight of more years. There may be a trace of sadness in her face. But she has placed her arm around the waist of the woman standing next to her; no one else in the picture has assumed this pose. Perhaps the gesture was dictated by the photographer, but I like to see it as a mark of sisterly affection toward her colleague on the resident faculty.

The twenty-five students who graduated from the Chicago Training School at the end of Amorette's first year came from thirteen states and from regions as distant as New York and Connecticut, to the east, and California, to the west. No graduate came from the south; the Methodist Church had split over the slavery issue in 1845 and was still divided, the Southern branch calling itself the Methodist Episcopal Church, South. The largest group of students, not surprisingly, came from Illinois, but only one came from Chicago. She was M. Jennie Birdsall. Amorette will meet her again, years later, in California, and this time Miss Birdsall will be supervising her.

All the graduates would have spent a year mastering a demanding course of study. Bible was taught an hour a day, five days a week. Two medical lectures a week were scheduled; even students who were not studying to be nurses were expected to learn the rudiments of care for the sick, since they would almost certainly encounter sick people as they made their home visits. Likewise, students who were preparing to be nurse-deaconesses needed to be well versed in Christian teaching, for they were expected to evangelize when they cared for their patients. Stories of the ways these expectations were met are written into the record of the deaconess movement.

Isabella Thoburn, a missionary returned from India who joined the first group of Training School deaconesses, told of a widow whose child had diphtheria. Forced to work all day and watch over her sick child all night, the poor woman was nearly exhausted, until a deaconess came to her home, took over the care of the child and allowed the mother to rest. "We cannot effectively carry the Gospel to such people," Thoburn wrote, "unless we take with it the love that is ready to help in time of need, whatever the need may be." Here, succinctly put, was the dual aim of the deaconess movement: to relieve the physical suffering of the poor and spread the gospel of Christ among them.

This approach was equally effective in Jewish homes, wrote Lucy Rider Meyer: "Our deaconess nurses find some cases of peculiar need among the Jews, and are able to gain an influence over them, after caring for them in sickness, that could be obtained in no other way. . . . It would be hard to conceive of a better way of breaking down prejudice and convincing our Jewish people that the Jesus they have been taught to abhor, but who can inspire his followers to such work, is the Messiah they have been expecting."

Reading the words of Rider Meyer today, we hear condescension and intolerance where she intended only zeal in the Lord's work. I have no idea whether Amorette shared the common prejudice against Jews and, unfortunately, no reason to believe she did not.

Some time during Amorette's tenure as matron, in the midst of her other duties, she satisfied the requirements that qualified her for the deaconess's license. *The Message,* the organization's newsletter, commemorates her accomplishment: "Mrs. A. E.

Foote, our good training school Matron, received at last recognition by the church for the labor of love she has so long been doing." Amorette was one of ten women who were consecrated as deaconesses in the Clark Street Methodist Church on the evening of November 12, 1891. *The Message* calls it "a solemn and beautiful service."I envision nine black-clad women standing in a row; the tenth is absent because she has contracted typhoid fever, probably from a patient she was nursing. The nine are listening intently while the charge is read to them:

"Dear Sisters. . . . You are to minister to the poor, visit the sick, pray with the dying, care for the orphan, seek the wandering, comfort the sorrowing, save the sinning, and relinquishing wholly all other pursuits, devote yourselves to such forms of Christian labor as may be suited to your abilities." While few of the enumerated duties apply directly to Amorette, the final one surely does. At her age, with her past experience, being an institutional mother is the form of Christian labor best suited to her abilities.

To some in the audience, the sight of these women pledging themselves to be handmaidens of Jesus Christ may have been an uncomfortable reminder of the resemblances between a deaconess's life and that of a Roman Catholic nun. Though it was always agreed that the deaconess's garb would not be "Romish," since it would conform in a conservative way to contemporary fashion, the fact remained that she would wear a sober black dress and distinctive black and white headgear. And while it was true that she would not take an explicit vow of chastity, she would have to remain single or give up her office. Widows were admitted, and deaconesses were not discouraged from marrying, but the phrase in the charge, "relinquishing wholly all other pursuits," made it clear that both marriage and deaconess work were full-time occupations.

Implicit in Amorette's understanding of this part of the charge was the knowledge that she would not be allowed to do the work she had been given to do unless she remained celibate. Her assent, I believe, solemnized a longstanding renunciation of sexuality. The records I have gathered strongly suggest that for most of her adult life sexual love was only a distant memory. She and Seth lived together for eight months before the Gold Rush separated them. Then he was absent for four years, then he was desperately ill, then he was gone. For the next twenty years or so she was occupied in looking after her son. Now she is seen entering an organization that excludes married women. Thus her widowhood becomes a part of her official identity.

There was one more way in which Amorette's life as a deaconess resembled that of a nun: deaconesses must expect to live in some kind of group home—in short, an institution. This restriction on a woman's private life, linked with the prohibition against marriage, gave rise to the criticism that the deaconess's life was an unnatural one for a woman. Isabella Thoburn offered a personal, and wistful, response: "The happiest place in the world is that where father and mother gather their children into a family circle, but the next happiest place is one like ours, where women to whom no other home has been allotted, may live and work and pray and rejoice together."

It saddens me to think that my grandmother, in late middle age, had become a woman to whom no other home was allotted. But I see the necessity of it. Amorette and Dell had made their home together longer, I imagine, than most mothers and sons did, even when the woman was a widow and the man an only son. But now Dell was independently, ambitiously establishing himself in his profession, as he had every right to do. Early in 1891 he traveled to Berlin and Vienna for postgraduate study in surgery and gynecology. Those months in centers of medical research expanded his knowledge and developed his surgical skills, which probably had not received a great deal of attention in the homeopathic medical college. Shortly after his return from Europe he got married, though I do not think he was entirely happy about it. That was another of the stories my mother told me. Before he left Omaha, so the story went, he had asked Milla Baird to marry him, but she had declined. Upon his return she told him she had changed her mind. He by this time had changed his mind too, but he felt obliged to honor his previous offer, and in September 1891 they were married. Two months later Amorette was consecrated as a deaconess.

No doubt it had seemed to both Dell and his mother that when he completed his medical education it was time for him to make a life for himself, by himself. That would explain why she took the job as matron but not why she committed herself to the austere life of a deaconess. Was it fear of loneliness? Did the prospect of life in an institution, with its assurance of constant companionship, attract rather than repel her? Was it the promise of financial security? Though she worked without pay, her living was provided, and she probably received the small allowance for personal expenses that was customarily given to deaconesses.

Any or all of these thoughts may have entered into her decision, but I believe that outweighing them all was a sense of religious mission: she wanted to work in the Lord's service. According to one student of the deaconess movement, most of the women who became deaconesses did so because they felt they were called to the work by God. I believe Amorette, like those others, felt the call. There may have been some ambition intermixed with her piety; she may have welcomed the opportunity to make a broader use of the domestic skills and wisdom she had developed over the years, and be recognized for them.

But every scrap of evidence I have collected about this period in her life tells me that she believed she was serving God in her deaconess work and that she rejoiced in that belief.

A portrait photograph of Amorette, taken at about this time, shows her looking thinner than she did in the photograph of the resident faculty, which was probably taken a year or two earlier. Possibly she had been ill; undoubtedly she had been working very hard. Nevertheless, she looks strong, assured and happy.

In the fall of 1891 Dell attended an evening meeting at the First Methodist Church of Omaha. Under consideration was a proposal to establish a Methodist hospital and deaconess home in that city. The proceedings were recorded in a book I discovered at the Sutro Library in San Francisco. I had gone to this small, rather quaint branch of the state library system by chance, simply because my husband wanted to stop there and look up a few things before we went on to a luncheon at San Francisco State University. As we were leaving home, Henry had said, "You might find something of interest to you at the Sutro," and so I took my packet of index cards with me.

Having the idea, at that time, that Amorette had done her deaconess work in Omaha, I began to look through the card catalogue, which the Sutro, in defiance of the Information Age, still clung to. What I discovered was a book called *Methodism in Omaha,* written by the Reverend James Haynes. I took it back to my table, where I soon found myself reading a laudatory description of Omaha's newly built First Methodist Church. This was the church where I, as a young child, had attended Sunday school. When the Reverend Haynes came to describe the side entrance that led to the Sunday school rooms, I remembered that entrance and one of those rooms, the one with the picture of a little girl and a robin on the wall, and on a table below it a round box in the shape of a cake, with a slot in the lid where you dropped your birthday pennies.

I turned to the chapter on the establishment of the hospital and deaconess home. When I saw my father's name in it I was not surprised, for I had always understood that he was one of the founders of the hospital. The surprise came when I read what he said at a meeting of the organizing committee—he was talking about his mother. "Dr. D. A. Foote," wrote Haynes, "having lately come to the city from Chicago, whose mother was connected with the training school of that city, was called upon for some statements. He was familiar with the work, and gave some facts corroborating the points made by the others; and which, together, made such an impression as to awaken stirring interest." From this unexpected source I learned that I should stop looking for Amorette in Omaha and start looking for her in Chicago. And I learned that her work, and my father's pride in it, had helped to bring a deaconess program into existence in Omaha. A more direct influence, I should add, must have been the talk by Lucy Rider Meyer, who was the principal speaker at the meeting that night.

Before the year 1891 was over, Rider Meyer and her husband, with their young son, moved their residence out of the Training School and into a cottage in the suburb of Oak Park. Their move would seem to have made Amorette's role of institutional mother all the more crucial to the school's operation, but she evidently was preparing for a move of her own. The spring semester, 1892, was her last in Chicago.

That summer she went to Omaha to stay with Dell and his wife, who were expecting their first child. At the end of the summer she came through Chicago and paid a visit to the Training School. She was on her way to a new job. *The Message* wrote, "Mrs. Foote, who certainly deserves the first place in our list notwithstanding some one facetiously declares she should go to the foot, passed through Chicago on her way to her new position as Matron of a school for colored girls in Alabama. She is very well, and hopes her stay in the South will permanently benefit her health."

I said earlier that Amorette entered into her professional career when she took the job in Chicago. My reasoning was that although she was not paid for what she did, her work called upon her best knowledge and skills, and she engaged in it, full-time, for eighteen years. Yet *profession* is surely not the word Amorette herself would have chosen. Like her sister deaconesses she would have said she had found her calling. Now, as I seek to place her in the broader context of nineteenth-century women's history, I prefer to say she found her vocation. The Latinate form sounds less ecclesiastical—more, well, professional. But of course it means exactly the same thing.

Foster Hall, about 1895. Courtesy of Talladega College Archives, Talladega, Alabama

Chapter 20

The Matron of Foster Hall

On a soft gray afternoon in October I was walking up a long sidewalk that led to a three-story red brick building. The light rain that had fallen on this small Alabama town in the morning had left damp patches on the squares of concrete and beaded the grass with water-drops. Beside me walked Dr. Helen Bush Caver, librarian at Talladega College, a tall black woman, imposing in carriage—a strong woman, with a quick mind, a hearty laugh and an obvious devotion to the college. Dr. Bush Caver, having spent much of her morning supplying me with documents from the library's archives, was now taking me to the place where Amorette had worked for seven years of her life. From a distance Foster Hall was still handsome, the rectangular plainness of its front softened by the gables that rose at either end. But when we came up close I saw the peeling paint, once white, on the supports of the front porch. Many of the windows were cracked or broken.

I looked up at the letters on the brick facade, letters that had formerly spelled out the building's name, and I saw that the F had fallen over and hung upside down, while the O was missing altogether.

We mounted the porch steps. Pointing to the lefthand windows on the first floor, Dr. Bush Caver said, "That's where the house-mother had her apartment when I was a student here."

The double doors of the front entrance were padlocked but not tightly shut, so that I could look inside through the space between them. It was unlighted of course, but I dimly made out a wide central hall. In the middle of it a typist's chair sat by itself, looking as if it had rolled there of its own volition. On either side I saw a table or two, a file cabinet or two—all of them having given up all pretense of doing useful work.

I turned away from the door, knowing I had come as close as is now possible to entering the world my grandmother inhabited. Exactly how she arrived there I can only imagine.

Riding from Chicago to Talladega, she has to change trains more than once. By this time in her life she has become quite accustomed to train travel, in the middle west, that is. She knows the route from Chicago to Omaha very well, but she has never been to the South. The long train ride gives her an opportunity to think about the duties that lie ahead for her. She is an experienced matron, but the girls under her supervision this time will be younger than the students in Chicago.

Even more daunting, they will be "colored"—a term regularly used by refined white people like herself to denote persons of African descent. In her fifty-three years of life Amorette has had little contact with colored people. True, there was the enclave at Frog Hollow, settled by former slaves, and some of those people had come to Fayette to work. That was where

Susan Collins had come from. But there is no reason to think that Amorette knew Susan Collins at this time.

Through Illinois, Kentucky, Tennessee. Over rivers: the Ohio, the Tennessee, the Tallapoosa. As the landscape grows more strange to her, the questions about her new position grow more insistent: Will the missionary spirit which she knows she possesses enable her to get along with her colored girls? Can she discipline them when they need discipline? Will they respect her? Will they obey her? Can she come to love them as she had loved the white women in Chicago?

The buggy that takes her from the railroad depot to the college campus comes to a stop before a walkway lined on either side with trees. Looking up its length, Amorette sees an elegant brick building, dignified yet welcoming. A front porch with white railings stretches out on either side of the entrance; slender wooden columns, also white, support the porch roof. Rows of tall, many-paned windows, each one crowned with an arch of bricks, give the facade a classic beauty and promise an abundance of light and air inside.

Foster Hall was the women's dormitory; the men's dormitory was named Stone Hall. Clearly, Talladega College was not a school for girls only, as the item in *The Message* had indicated. Founded shortly after the Civil War by William Savery and Thomas Tarrant, two former slaves, it was designed from the beginning to educate both the sons and the daughters of freed men and women. Very soon the cabin that was built to house it proved too small for the numbers of young people who came to be taught. Fortuitously, a Baptist high school built for white boys came up for sale. It was an impressive structure in the ante-bellum style, its massive white columns supporting an ornate pediment. Lacking the money to buy it, Savery and other leaders in the black population sought aid from General Wager Swayne of the Freedmen's Bureau, who in turn asked for help from the American Missionary Association. Contributions of money from both organizations made the purchase possible, and the AMA added Talladega to the list of Southern schools it directed. In November 1867, 140 students, most of them only a few years out of slavery, passed between the tall columns into the front entrance of Swayne Hall, to face duties and opportunities previously reserved for whites. The irony of the sight cannot have been lost on those who watched. To the black citizens of Talladega it was cause for rejoicing; the white citizens, by and large, saw it as an insult.

The new institution identified itself as a college, though its officers were aware that the students who came to them were not yet ready to do collegiate work. But the word expressed a hope and a goal. Talladega was the first college in Alabama to be open to colored people. White students were not excluded—American Missionary Association schools were open to all—but almost all Talladega students were black and almost all its teachers were white.

It is not clear how much of this history Amorette knew when she walked through the double doors at Foster Hall for the first time. It is even possible that she herself was the source of the error in *The Message.* Certainly she was unwise to have expressed the hope that her new position in the South would benefit her health. The American Missionary Association had from its inception minced no words on that point, warning applicants that "this is not a hygienic association, to help invalids try a change of air." Health was one of the six qualifications set forth in a circular which the AMA sent to applicants for teaching positions. The first was "missionary spirit," a willingness to endure hard conditions "as a good soldier of Jesus Christ."

Another was "experience"—in particular, successful experience as a disciplinarian.

A year after the first six were circulated, one more condition of employment was added. This one, proposed by AMA Treasurer Lewis Tappan, was aimed exclusively at white people who worked or hoped to work in AMA schools. If they could not "'conquer their prejudices,'" Tappan wrote, they "'evince indisputably that they are unfit for the high and responsible office of superintendent or teacher, and should be dismissed from the post they so unworthily fill.'" Tappan's pronouncement, which first appeared in the AMA's monthly journal, was subsequently written into a resolution that became permanent AMA policy. In its final form, the resolution spoke explicitly of prejudice "against the people of color."

Assisting the people of color was the reason for the AMA's existence: it was founded as an abolitionist society. And after emancipation, writes historian Clara De Boer, the AMA "did more to educate the freed slaves than any other organization, including the Freedmen's Bureau." Though its founders were Protestant clergymen—many of them Congregationalists—the schools they established accepted students of all religions, or of no religion. Once students were enrolled, of course, their education was shaped by Christian doctrine and Christian practice.

Into this rarefied atmosphere came my grandmother, a Methodist among the Congregationalists, a middle-aged white woman, over-concerned perhaps about her health, pledged to meet hard trials with Christian endurance. Entering the long, wide central hall, she could see the doorways of various public rooms. The most used was the dining room, where both students and teachers (those members of the "lady faculty" who lived there) ate their meals. The kitchen, however, was in the basement, meaning that all the cooked food would have to be carried up a flight of stairs before it was served.

There were parlors on the first floor, and one of them contained a piano which students were allowed to play, after obtaining permission from the matron. There was also at least one smaller room that could be closed off for faculty meetings. Privacy, however, was not much in evidence in Foster Hall. The sleeping rooms, most of them on the two upper floors, were designed to accommodate two students per room. Perhaps the teachers too were asked to share rooms. Nevertheless I think—I hope—that Amorette had a room of her own. Encouraged by Dr. Bush Caver's recollection, I am placing her in that first-floor room just inside the front door, where she could keep an eye on things.

Less than two months after her arrival in Alabama she was faced with a serious disciplinary matter. A student named Cora Jeffries was suspected of stealing from other residents in Foster Hall. The complaints against Cora were very likely brought first to Amorette, since she was the faculty member always on duty there. She in turn would have notified Miss Alice Topping, the preceptress. The two of them would then have taken the case to the Prudential Committee, of which both women were members.

At the Committee's November 4 meeting, the nine members heard with concern a growing number of complaints against Cora. A student, Hattie Foster, believed Cora had taken a dollar from her. Professor T. N. Chase, who had originally recommended Cora's admission to the college, revealed that he had suspected her of dishonesty for some time. Then came an accusation of a different kind of theft, a kind particularly distasteful in an academic setting. Lilly Milhous, another student, complained that Cora had stolen an essay Lilly had written and had read it aloud in class as if it were her own. Cora, when confronted, denied the charges. Unwilling to discipline her without conclusive evidence, the Committee gave the task of investigating the complaints to three of its members: Miss Topping, Professor Martin Lovering and Edwin Silsby, the college treasurer. Three days later, after listening to the group's findings and discussing them, the full committee voted that Henry De Forest, the college president, should reprimand Cora and tell her that she "must exhibit perfect behavior" and "earnestness in study" if she hoped to continue in school.

But Cora was not content to let the matter rest. A week later she began to claim that she herself had lost three dollars. Her claim was not believed. Those who knew her, including classmates and friends, thought she had hidden the money so that she could pretend it was stolen. In an effort to get at the truth, Miss Topping asked Cora to account for all the money she had possessed since she arrived in September and all the expenditures she had made since then. When the two columns were totaled, Cora's expenditures exceeded the amount of money she said she had. Faced with these figures, she confessed that she had taken Hattie Foster's dollar. With her confession as evidence of misconduct, the Prudential Committee voted to suspend Cora "indefinitely." She was to be "kept in her room" until her mother could be heard from. It would have been Amorette's job to provide Cora with food during her confinement. While she would probably not have carried trays to Cora's room herself, she would have given the orders that her student helpers would follow.

Six days before Christmas, at a regular meeting of the Prudential Committee, Amorette was appointed to a small subcommittee; it was the first time she had been given this kind of responsibility. A problem had arisen concerning "the transient visitor abuse," and President De Forest asked Amorette, Miss Topping and Mr. Silsby to remedy it.

The name of Miss Topping, preceptress, headed the list of women faculty on the official roster for 1892-93. Amorette's name came last; that of Miss Pingree, matron of Stone Hall, next to last. But I do not mean to complain about my grandmother's low status in the Talladega College hierarchy. It was only logical that in an educational institution the women whose jobs were largely domestic in nature should be ranked below those who taught academic subjects. And indeed the names of all the women faculty members were listed below the men's, yet we should not conclude that the women's work was any less important in the day-to-day operations of the college. During the years Amorette worked there, women faculty members outnumbered the men by nearly two to one.

Edwin Silsby, the college treasurer, was notable among the faculty for establishing close relations with the black population of the town. As historians Maxine D. Jones and Joe M. Richardson have remarked, "Silsby spent more time in the black community than any other Talladega official, visiting blacks in their homes and churches and attending agricultural and educational meetings." President De Forest, who before coming to Talladega had been pastor of a Congregational church in Des Moines, Iowa, gave the black people gifts at holidays, and they were much appreciated. Still there was an air of charity about them. Silsby met black citizens on a person-to-person basis.

The nature of the "transient visitor abuse" was not explained in the Prudential Commitee's minutes. From the sound of it, one might think that the transient visitors were the ones being abused, but surely that was not the problem. I am guessing that a few young people who were not students came to visit friends in the two residence halls and stayed too long. Evidently the subcommittee's remedy, whatever it was, solved the problem, for the minutes never mention it again.

The year 1893 began badly for the country and for the college. In February a financial panic and the depression that subsequently settled upon the nation worsened the already serious poverty of the farming region served by Talladega College. Parents found themselves unable to pay the very modest tuition, so that even as more young people were enrolling in school, some of the best and most advanced students were dropping out for lack of money. Moreover the widespread hard times meant a loss of financial support from people who in the past had been donors. Yet there was another source of hope: this was a religious college after all. For two weeks in March, revival services called upon listeners to give their lives to Christ. With its preaching and singing and testimonials, the revival strengthened the faith of believers and converted most of the doubters, at least temporarily. Obviously gratified, Dr. George W. Andrews, dean of the department of theology, reported on the revival's success: "The great majority of our students in Stone and Foster Halls were professing Christians before the meetings began," he wrote, "and now that they have closed I know only four or five who do not avowedly belong to the Lord."

In the face of the deepening depression that spring, the Prudential Committee did what it could to make student life more agreeable. In early May its subcommittee on entertainment extended the privilege of making social calls at Foster Hall "to all young men boarders who belong to the Senior—Normal, College, and Theological classes." Obviously the resolution covered a limited number of men. "Boarders" meant only those who roomed in Stone Hall and took their meals in the Foster

Hall dining room. The Normal class was composed of students who were preparing themselves to be teachers. "College" meant those who were headed for a bachelor's degree, which Talladega only now, some twenty-five years after its founding, was able to offer. The Theological class was confined to the small number of superior students who had chosen to specialize in that discipline, most with the intention of becoming ministers. And of course the word *Senior* excluded any young men not in the final year of their studies.

This select group would be allowed to make social calls at Foster Hall "on Saturdays between the hour of 4 o'clock and the ringing of the first supper bell." Now Amorette had a new duty—to keep an eye on the young couples sitting in the parlor or strolling about the yard on Saturday afternoons. It was a busy time of day for her, with supper to be seen to, yet the sight of her girls, happy in their new privilege, must have been a pleasant one.

June. . . commencement week has ended, Foster Hall has emptied for the summer months. Amorette is returning to Omaha, which she is now calling home. She will spend the summer with Dell and Milla, as she has been doing since their marriage. Perhaps this summer she can give Milla a hand with the care of baby Marjorie. Babies need special care in the summer; it was in the summer that Adelia's baby died, so many years ago. But times are different now; doctors know more about babies' illnesses and how to prevent them. Dell is a doctor—surely he will keep his daughter safe and well.

. . . Marjorie, under the loving care of her parents and her grandmother, thrives and grows. In the fall, about the time the fierce sun of a Nebraska August gives way to the mellower gold of September, Amorette boards the train for Alabama, where summer is due to last another month at least.

When she returned to Talladega College, she found that she was to work under a new preceptress, Miss Mary Strong. Miss Pingree was still matron of Stone Hall. The Prudential Committee resumed its meetings, and the three women, along with President De Forest, formed a subcommittee to work on the Cards of Rules. These were two single pieces of cardboard, "Rules for Young Women" and "Rules for Young Men," which were probably designed to be posted in Foster and Stone halls. The young women's rules call for a self-discipline that must have been difficult for some of these adolescents to maintain: "7-9 p.m. Students are required to be in their rooms before the tardy bell rings, and are not to leave them during study hours, during which period no disturbance is allowed. . . Students are required to be in bed, and the lights extinguished at the last bell. Between the last bell and 5:30 o'clock in the morning, no talking or leaving rooms is allowed."

The rule card does not allow for emergencies, such as a sudden illness, which might make it necessary for a girl to leave her room to summon aid for her roommate. It is very specific, however, as to proscriptions: "Young women will avoid . . . unnecessary noise, uncleanness, improper language and immoral tendencies."

More and more, it seemed, Amorette's work revolved around the making and enforcing of rules. The "food from home" rule, discussed at a November meeting of the Prudential Committee, surely fell within her sphere of responsibility. The college catalogue had told students in plain words that while they were at school they should not receive gifts of food from home. Still the gifts kept coming, as Miss Pingree and Mrs. Foote could not fail to see. It was time, the Committee decided, to go to the source of the food. They agreed that postcards should be sent to all parents, requesting them "not to send eatables to their children."

What was the college's reasoning here? Why, in a time of economic depression, did it forbid those gifts of food? Perhaps it was a question of fairness; students whose parents were very poor might have to subsist on less food than their slightly more prosperous classmates. It might have been a simple matter of housekeeping, since food kept in students' rooms would attract rats, mice, roaches or any other pests that might have found their way into Foster Hall. Or perhaps it was a continuation of an attempt, begun in the 1880s, to wean students away from the not-very-nutritious diet—Mr. Silsby called it the "hog and hominy" fare—which they ate at home. Possibly it was a combination of all three.

Keeping her girls in line was a taxing job for Amorette, but her work was not unappreciated, at least in her own family. Dell, for one, was proud of her. That fall, in a letter to his Uncle David Foote, he must have described the work she was doing, for David replied with words of praise. David was Seth's brother, youngest son of Moses Foot's first marriage. Like Dell a physician, he lived in Schoolcraft, Michigan. Moses, now eighty-one, lived with him. Moses sent a message to Dell, by way of his son: "Father says tell you he has six Grand children & six great Grand children." Moses also helped David compile the list of Foote ancestors and descendants that Dell had requested. I note that David too has added the *e* to the family name.

The genealogical information that David enclosed covers several pages, and my father was undoubtedly glad to get it, but two other portions of David's letter are of greater interest to me. First, it is gratifying to see that Moses is on good terms with his grandson Dell, the old tensions between Moses and Amorette's family evidently forgotten. Even more gratifying is David's obvious appreciation of Amorette's work and her character: "I must say your Mother's zeal is very commendable may God Bless her labors & supplement her efforts in doing good."

Zeal: a determination to do good work, the Lord's work, that is what we see in Amorette at this time, in this place. And her circumstances were exceptionally difficult just then, for an atmosphere of racial violence pervaded the state of Alabama in 1893. At the end of the year, when the number of lynchings, state by state, was tallied, Alabama headed the list. As Ida B. Wells reported in *A Red Record,* twenty-five colored persons had been lynched in Alabama that year. The town of Talladega and the college were spared; nevertheless they lived with the fearful knowledge.

As a way, perhaps, of asserting its right to a normal existence despite the troubled times, the college resolved to maintain its women's dormitory in the manner it clearly deserved. In March of 1894 the Prudential Committee voted that certain repairs to Foster Hall were "of first importance and necessity." Looking down the list of those important and necessary items—"kalsomining girls' rooms; oiling the floor; new roof; painting woodwork in parlor; sundry improvements in dining room and kitchen; improvements in walks; painting outside woodwork"—I conclude that Amorette must have had a good deal to do with drawing it up. These were the things she would know about, firsthand. Kalsomining [whitewashing] the girls' rooms would brighten their walls, sundry improvements in the kitchen and dining room (whatever they were) would surely lighten the girls' labor, and her own. Painting the woodwork in the parlor would make social hours more pleasant; repairing the roof would keep out the rain. Taken together, these repairs would lift the spirits of the residents of Foster Hall, if the money could be found to pay for them.

In the middle of May, Amorette is pleased to learn that Dell and Milla have had another baby, a son. They name him Arthur, the middle name Dell had taken when he discarded the original *Angelo.* Now, when Amorette goes to Omaha for the summer, she will have this child to care for—to bathe, to dress, to rock. Tending to the needs of a baby boy had been Amorette's initiation into adulthood, her solace in her loneliness. Lately she has dealt mainly with girls, but she will remember what a boy is like, once she holds him in her arms.

Soon after she returned to school in the fall, the Prudential Committee found it necessary to clarify the question of permissions. Sometimes students wished to leave their residence halls or the grounds around them, always they had to apply for permission to do so. The Foster Hall girls were required to ask the preceptress; the boys of Stone Hall applied to Miss Pingree. But one group of male students did not live in Stone Hall; they roomed in the barn and did useful work for Amorette, cultivating the vegetable garden and the orchard, milking the cows, gathering the eggs. To whom should they apply for permission to be temporarily absent? The Committee gave the logical answer: to Mrs. Foote.

Chapter 21

Matters of Right and Wrong

The boys who slept in the barn were put to a severe test in early February 1895, when Talladega suffered the coldest weather ever recorded for the region. Snow began to fall on Thursday the 7th; that evening a blizzard struck. The next morning the temperature at one citizen's home registered zero; at another's, two degrees above. Temperatures began to rise on Saturday, and on Sunday night rain fell, but it turned to sleet. On Monday morning the 11th the ground was "covered with a sheet of ice," as the local paper reported, and icicles hung from the trees.

How the girls of Foster Hall, and the teachers and the matron, shivered through those cold nights is difficult to imagine. I do not know of any heat source for the upstairs rooms. I doubt if many of the residents had enough blankets, since weather this cold could not have been expected. During those days of snow, rain and ice Amorette must have shuddered to see her girls come back from class cold and dripping wet. For on that same Monday, prodded I'm guessing by Amorette, the Prudential Committee voted that "every girl boarding in Foster Hall shall possess umbrella, waterproof and overshoes." The new regulation, to be printed in next year's college catalogue, did nothing to help this year's girls, but it was the best the Committee could do.

Before the month was over, a storm of another kind struck the Talladega campus, and Foster Hall lay directly in its path. At 7:30 on the evening of February 28, a Thursday, Amorette took her seat in the meeting room with the other members of the Prudential Committee. It was not a regularly scheduled time for them to meet. President De Forest had called them together to deal with an unexpected defiance of school authority.

Amorette already knew, probably better than most of the other members, what circumstances had given rise to this meeting. That morning, in Foster Hall, a group of her girls had banded together to protest the disciplinary treatment imposed upon a girl named Irene Waller. Irene was being kept in her room as a punishment, and her friends came to believe that she was not being given enough to eat. More than twenty girls crowded into Emma Ratcliffe's room to listen as two or three of the most articulate outlined a plan of action: they should all refuse to go to class until they could be assured that Irene was receiving adequate food.

These were the facts of the rebellion as Amorette understood them. Naturally she knew all the participants, and she may have felt that she herself was the most immediate target of their protest, since she would have been the person in charge of the food served in Foster Hall. The restrictions on Irene's diet probably came from a higher authority, such as the preceptress, but it must have been Amorette who translated them into tangible, edible form.

When all the members of the Prudential Committee had assembled, they bowed their heads as they were led in prayer. Then they heard the story of the protest meeting and its anticlimactic outcome. The resolve of most of the girls had quickly weakened; they did go to their classes, though some came in late. In the end three girls were identified as the ringleaders: Emma, who had allowed her room to be used as a meeting place, Adifi Sullivan and Georgia Patterson. The three were brought before the Committee. After questioning them, the members discussed the disciplinary action appropriate to each.

Whatever Adifi said to the Committee was enough to convince them that she had not been a ringleader after all, and no punishment was laid down for her at that time. But Emma and Georgia did not get off so lightly. They were prohibited from attending any school exercises, they were not to visit other girls in their rooms or receive visits from them and they were not to leave Foster Hall yard without permission. They were, however, allowed and in fact required to continue doing their assigned work in the Hall. In the hope that the two girls might be "brought to a proper realization of their wrong conduct, and to repent for it," Dr. Andrews and Mr. Martin Lovering were asked to have private talks with Georgia and Emma the next day. Then the meeting was adjourned until the following night.

Late as it was, President De Forest decided to talk with the two girls himself before he went home. What Georgia said to him that night so shocked and angered him that by the end of the interview he had made up his mind that she should be dismissed from the college. He did not share this opinion with the Prudential Committee until at least five days had gone by, but it obviously influenced the recommendations he made in the meantime.

When the Committee assembled on Friday evening, Mr. Lovering and Dr. Andrews reported on their interviews with the two girls. Mr. Lovering spoke first, admitting that although he had tried to make Emma see "the wrong of her conduct," he had not succeeded. Dr. Andrews then read his report, which he had written in advance: "'Georgia resolutely and persistently affirmed that her only thought was that Irene did not have proper food, so that she felt she did right in trying to secure it for her; and just as resolutely and persistently declared that in deciding not to go to her recitations she had no thought of interfering with school discipline.'" When he had finished, he turned over his paper to William Hutchison, secretary, who would copy it into the minutes.

Mr. Hutchison always wrote his minutes in a clear, legible hand, for which later readers can be very grateful, but in recording the story of this student protest, he left us with some unanswered questions. We do not know what the offense was for which Irene Waller was confined to her room. Up until this time, evidently, she had been a promising student, honored by being appointed Local Editor and Society Reporter for the *Talladega College Record.* We might speculate that she had made some injudicious remark in the paper, but of course we do not know. Nor do we know what food, nor how much, was being brought to her room for her. Nor whether kept in her room really meant locked.

After discussing the two men's reports, the Committee voted unanimously that Georgia Patterson, on account of her "persistent insubordination and insolence,"should be suspended at least until the end of the present school year. Once they made that decision, they postponed consideration of Emma's case, as well as the cases of the other twenty-three girls. On Saturday morning, March 2, President De Forest interviewed Georgia again. He told her that the faculty had suspended her and that she was to be sent home that very day.

As things turned out, she was not sent home on Saturday, but by Monday the president had decided he could tolerate no further delay. That morning he circulated a memorandum to all the members of the Prudential Committee. Describing his latest interview with Georgia, he wrote,"'Her contumacy was phenomenal.'" (*Contumacy*—a word that may have sent my grandmother to the dictionary, as it did me. It means, I discovered, "wilful and obstinate resistance to authority.") By the time he wrote the memorandum a report had reached him that Georgia "now regrets her action, and affirms that if she was disrespectful she did not mean to be." The president made no comment on Georgia's avowed change of heart, apparently questioning its sincerity.

His memorandum, which had to be hand-carried from one Committee member to another, explained that it was not convenient to call the members together before the train was scheduled to leave for Selma—evidently Georgia's home. Therefore the president asked each member to answer, in writing, the following question: "'Shall she go home today?'" He attached a separate sheet of paper, on which members were to record their votes and any comments they wished to make.

The first person to receive the memorandum was Amorette. She wrote "Yes" and sent the papers on to Miss Jane Ainsworth, who was the preceptress that year. She also answered in the affirmative. Other members followed suit until the memorandum reached Edwin Silsby. The report of Georgia's contrition had evidently moved him, for he dissented, saying, "'I am willing to accept a suitable apology if offered and hold on probation.'"

When the memorandum reached Dr. Andrews, he equivocated, voting "Yes" but implying that he did so only to go along with the majority. Last to vote was President De Forest, who added his assent to the list. Mr. Silsby's was the only "No."

I am disappointed in my grandmother. I like to think that if I had been in her place I would have given Georgia another chance. But I understand Amorette's position. Upholding authority, enforcing the rules—that was what she was hired for. And I do not mean to imply that her "Yes" vote was motivated by expedience. I think she voted the way she did out of conviction. I think she saw it as her Christian duty to support the president and maintain discipline.

Before the day was over, Georgia was on the train bound for Selma. Somehow, without being seen and stopped, six male students boarded that same train and accompanied her part of the way, to give her comfort and show their support. They got off at Barclay's Station, a train stop a few miles out of town.

When the Committee reconvened on Tuesday, President De Forest made his recommendations for the treatment of the remaining twenty-four participants in the protest: Five girls—Adifi Sullivan was one of them—were put on special probation. Nineteen girls, listed by name, were thought not to have understood the seriousness of their action; they were "admonished and reproved" and "restored to their places in school." Emma Ratcliffe was on this list. All the Committee members assented to this recommendation, even Mr. Silsby, but in casting his "yes" vote he pointed out the injustice of the severe punishment imposed on Georgia. "'I vote yes,'" he wrote, "'but think this action inconsistent with the majority decision concerning Georgia Patterson.'" Dr. Andrews indicated that he too saw the inconsistency.

The case was still not closed; the Prudential Committee

would have to meet twice more that week. A day or two after the Tuesday meeting they learned about the young men who took the unauthorized train ride. On Thursday evening the Committee questioned each of the offenders separately and after hearing all six, voted that they were "'hereby reproved and admonished.'" It was the same mild penalty the nineteen girls had received.

That night the Committee asked President De Forest to prepare a "circumstantial report," in writing, of his two interviews with Georgia. He had his report ready when the Saturday meeting opened. All members were present; all bowed their heads as usual while the prayer was offered, and all listened as President De Forest took up his paper and began reading. First he recounted the things Georgia said to him on the evening of the protest, making it clear that he had no quarrel with her manner toward him. She "expressed a regard for me personally," he said, "and as far as I am concerned, was entirely respectful." The things she said about other faculty members, however, aroused his indignation. "'She did mention others of our corps by name, speaking of them disrespectfully and very bitterly.'"

If I had been Amorette, I would have winced when I heard the words "others of our corps." He did not divulge the names publicly, so that Amorette could not know whether Georgia had named her. But because of the nature of her job Amorette saw Georgia every day; they slept under the same roof and ate their meals in the same dining room. Moreover, she was the one who was enforcing Irene Waller's punishment. I cannot help suspecting that Amorette was one of the objects of Georgia's disrespect and bitterness.

"She said," the president continued, "that we walked over the students because they were colored, or that we trampled on them, and twice made this affirmation." It seemed important to him that she said it twice, and so it was. To say it once might have been put down to a sudden, irrational burst of anger. Saying it twice suggests that she was convinced it was true.

Georgia was making this accusation against white men and women who had come to Talladega, often at great personal sacrifice, in order to help young people like her, to lift them out of poverty and ignorance. No wonder the white people were indignant. No wonder they called her insolent and insubordinate. But what if there was truth in Georgia's outburst? What if some of the faculty, despite their good intentions, did deny the black students their full humanity? Where should one draw the line between the imposition of discipline, surely a necessity in any school, and the enactment—in cold words, in a stiffening of facial expression, in a shrinking away from personal contact—of an only partially suppressed racial prejudice?

This was the danger Lewis Tappan had foreseen almost thirty years earlier. Teachers or superintendents who could not "conquer their prejudices," Tappan wrote, "should be dismissed from the post they so unworthily fill." Shortly after Tappan's unequivocal statement was published, the principle it embodied was incorporated into the official policy of the American Missionary Association.

Georgia probably had no knowledge of the AMA's stated position on racial prejudice, but rightly or wrongly she sensed its presence at Talladega. In this first interview with the president, an occasion when she may have felt encouraged to speak the truth as she saw it, she told him things he did not want to hear. Her words, and the passion with which she spoke them, convinced him at once that she did not belong in Talladega College. "It seemed to me at the end of that interview," he wrote, "that she ought to have been removed from the school, and that it was a great mistake to have tolerated such a girl so long." Clearly implied in his statement is a judgment that she ought to have been removed before this because she is a girl of bad character.

Reading on, the president described his second interview with Georgia, which took place on Saturday. "First, I carefully asked her how her conduct, after many hours for reflection seemed to her. She expressed no regret whatever. There was not the slightest evidence of penitence. When I told her that the Faculty had suspended her and that she was to go home that day, she impudently and defiantly said, 'I won't go!' I replied that she would go, if she was carried out of that Hall by main force."

The president has kept his angry reply in his memory for a week; now he quotes it verbatim to his faculty. It sounds as if he were proud of it. We can understand his feelings, perhaps even empathize with them, when we consider that he has worked for sixteen years to strengthen the college and elevate its reputation. Furthermore, those of us who have been teachers know how infuriating it can be when a student refuses to obey us. Yet his indifference—and the indifference of the faculty as a whole— to Georgia's complaints suggest that some of them had lost sight of the Tappan principle.

From his first interview with Georgia, evidently, the president's conviction that she must be removed from school never faltered. When he received the report that she regretted her action, he disregarded it. Henry De Forest was a minister of the gospel before he became a college president, but in this instance Mr. Silsby, a layman, seems to have had a clearer understanding of Christian charity.

And so, four days after the short-lived rebellion, Georgia was removed from Foster Hall (though probably not by main force). The crisis that began when a girl thought she was doing a kindness to a friend was over. The other girls who participated in the protest were warned by her example. Irene Waller, who, perhaps inadvertently, had started the uproar, lost her job on the *Talladega College Record.* By April another girl's name appeared on the masthead as Local Editor and Society Reporter. Discipline had been maintained, authority upheld.

Chapter 22

My Student Protest

One reason I sympathize so strongly with Georgia Patterson is that when I was in college I led a student protest myself. I was a sophomore at Grinnell College, and the country had been at war for nearly a year. At issue were college policy, army policy and attitudes regarding white girls and "Negro" (the word we used then) soldiers.

On the subject of relations between black people and whites, Grinnell had reason to be proud of its history. It was founded by a group of young Congregational missionaries, the "Iowa Band," who came out from New England before the Civil War to serve churches, save souls and establish a college. They chose Davenport as the site for the new school, which they named Iowa College. It proved not to be a happy choice. So fervently did Iowa College preach both abolitionism and temperance that citizens of the Mississippi River town made it feel decidedly unwelcome. In December 1858, responding to an invitation from Josiah B. Grinnell, Iowa College moved from Davenport to the prairie village that bore his name. Two months later Josiah Grinnell welcomed the notorious John Brown and a party of fugitive slaves to town. Defying the Fugitive Slave Law, Grinnell allowed Brown's runaways to sleep for two nights in his wool-storage barn. In time, Iowa College changed its name to that of Grinnell, the radical abolitionist.

My classmates and I had learned these stories as part of our college's heritage. When, in the fall of 1942, we heard that an Officer Candidate School had been established on the Grinnell campus, and that some of the soldiers were Negroes, we accepted the arrangement without noticeable surprise or complaint. Turning over the Alumni Recitation Hall to the Army meant that we had to attend class in some odd places (the basement of the freshman women's dormitory, for example), but such inconvenience was slight compared to what the soldiers would face when they left us, and I think we realized that. While they were there, we wanted to make them feel welcome.

Across the campus the welcoming hand was extended. The dean of the chapel invited the soldiers to college church. An editorial on the front page of the student newspaper, the *Grinnell Scarlet and Black,* told them, "We want you to feel as much a part of the college as your duties will permit." To help the soldiers mingle with us students and enjoy some part of college life, the women's social committee planned an open house "for officers, soldiers and college men." It was to be an evening affair; there would be "refreshments, games and dancing."

Dancing. The word called up the image of a brown hand placed in the middle of a white girl's back, in the prescribed position for social dancing, and a white hand resting on the brown man's shoulder. That image, or something like it, must have been what prompted college authorities to rule that only the white soldiers would be permitted to attend the open house.

When I heard that the Negro soldiers were to be excluded, I thought it wasn't fair. It was partly a simple matter of good manners, it seemed to me. If we were inviting our country's soldiers to a party, we should invite all of them. But it was also a matter of social justice. This was the time when my eyes had been newly opened to the shameful injustice of America's treatment of its Negro citizens. Professor Bernard Schilling, my mentor at the time, made me see and feel suffering to which I had paid little attention before. That year he had recruited a small group of students for his new Humanities major; we would meet from time to time at his house. On the night when we were discussing *Othello,* Mr. Schilling made Shakespeare's play the vehicle for a passionate indictment of racial prejudice. I hear him now, castigating all those who would persecute a man "on account of a condition he had not created and could not change." His words lighted a fire in my heart. I burned to do something that would help right this wrong.

I had already been made aware of the poisonous effects of racial prejudice. My mother had a song she sang to me when I was barely old enough to remember. It was about a colored child who cried because she (I thought of the child as a girl) was kept in her own back yard, fenced off from the white children who called her names. I felt sorry for the little colored child, as Mother perhaps had intended. Mother had also told me the true story of the lynching of a black man on the courthouse grounds, just half a block from my father's office. Daddy was keeping office hours, something he frequently did on Sunday afternoons, and Mother was serving as his nurse.

They began to hear a "horrible sound," she said, a kind of hum, rising to a roar, of thousands of maddened human voices. They knew something terrible was happening and they could tell where it was coming from, but all they could do was get away from that sound as quickly as they were able.

So I had been told what racial prejudice could lead to, and yet when I went to high school I seemed to forget the lessons I had learned at home. In the first semester of my freshman year I was in Webster Porter's Latin I class. Our school was on the top of a hill; Mr. Porter's classroom, looking east over the city and the river, felt like an aerie. Pigeons often alighted on the cement railing outside the window; sometimes Mr. Porter would lean out the window and talk to them. "Buckety-boo," he would say. "Buckety-boo."

There was a Negro boy in my class, an unusual thing in itself. Though Negro students were quite numerous at Central High School, I rarely saw them in my college preparatory classes. But this boy, with his unmistakably brown skin, was taking Latin. He and I were the best students in the class, or at least the most diligent and enthusiastic. I see him sitting in the center row of seats, raising his hand, smiling as he gives the right answer. In my picture of him he is wearing glasses, as I did.

When class was over he and I frequently chatted as we walked to the door, but after we crossed the threshold we parted, each of us plunging separately into the crowd. One day, however, he remained by my side, still talking, and we started down the hall together. Maybe he was talking about Latin; maybe he was laughing about Mr. Porter and his pigeons. But I became uneasy. I felt he was overstepping a boundary that I sensed, though I had never put it into words. I was afraid one of my friends would see me walking with a colored boy. I held back, turning in a direction I did not really want to go, just to get away from him. He must have realized what I was thinking, for he never tried to walk down the hall with me again.

Five years later, when Mr. Schilling taught his lesson on

racial prejudice, I did not remember the boy in my Latin class. I speak of him now to acknowledge that the girl who was swept away by an ardor to reform was also the girl who had been in need of reforming.

For I was swept away, and the party for the soldiers gave me an opportunity to be a ringleader in a protest against racial injustice. I was not the only ringleader. Some other girl, from some other residence hall, must have been the one who first told me about the plan for the protest. But I was an ardent recruiter. In the hope of persuading the authorities to change their minds, I went from room to room in my dormitory, urging the occupants to join me in a demonstration that night. They were a good-hearted group of young women, and a number of them responded to my call. The excitement spread through our residence halls with little difficulty, since they were all connected by a common passage. At its head stood Main Hall, which housed the dining room, the drawing room, some offices and the suite provided for the dean of women. Our plan was to gather on the lawn outside her suite and make our demand known.

By the time we were ready to take action it was full dark outside. I do not remember, probably never knew, exactly how large our numbers were that night. But I remember most clearly standing in the darkness among a crowd of girls, looking up at the dean's windows, where light glowed through drawn shades. Did we chant slogans or wave posters? I don't think so, we were too unorganized and too ladylike for such display. We were noisy, however, and the dean must have known we were there. But she did not show herself, and after standing around for a while we went back to our rooms, disheartened and feeling rather foolish.

Nevertheless it became clear the next day that we had made

our point. I was identified as a ringleader, and the first administrator to call me to account was the dean of women, Miss Evelyn Gardner. She was a tall, stately woman, more given to disapproval than to approval, in my encounters with her. Of course that was her job—to enforce the rules.

After reviewing the cause of our protest, Dean Gardner said to me coldly, "The Negro soldiers would not want to dance with you."

My immediate reaction was shame. To my ears she was imputing to me that grave offense against the accepted code of female conduct—running after men. The fact that the men in question were Negroes hinted at a shameful eagerness for a forbidden intimacy. I left the dean's office humiliated.

Then I was called into the office of the president of the college, Samuel Stevens, who smiled at me patronizingly and in the smoothest manner possible made me understand that as the holder of a generous scholarship I was expected not to challenge college suthority. That was the end of the interview and the end of any disciplinary consequences for me. A few days later the president said to Mr. Schilling with a laugh, "Well, if we will teach them to think for themselves, what can we expect?" As Mr. Schilling told me of the president's remark, he smiled too. I did not feel like smiling. It was painful enough to learn that my ardent campaign for racial justice had been dismissed by the president as a harmless folly. But to hear the man who had inspired my ardor now join the president in belittling it made me feel small and inconsequential.

At this point, obviously, the resemblance between Georgia Patterson's case and mine has weakened. Both our protests failed—the Negro soldiers were not invited to the party, Irene Waller was not, as far as anyone knows, given more food—but the consequences to me were slight. Since the president had not asked me to repent, I had no occasion to jeopardize my status in college by refusing. The only action I was able to take was to stay away from the party myself, and I did stay away.

Though I did not know it then, there was a cruel irony in the exclusion of the Negro soldiers at that particular time. That was because the inclusion of Negroes as officer candidates was such a new thing. In 1940 the Army had adopted a policy requiring that Negroes should be admitted to Officer Candidate School, and ruling further that they should attend the same schools the white soldiers attended. This was a radical departure in an army still almost universally segregated. The new policy, however, was slow to be recognized by the white officers responsible for recommending Negroes to the program. Nineteen forty-two was the first year in which the schools were graduating Negro officers in any substantial numbers. Grinnell's Negro soldiers, then, would have arrived at our campus heartened by the opportunity newly offered to them, only to have forced upon them another reminder of their inferior social standing.

I recently visited the Grinnell College archives to see if any trace of the exclusion of the Negro officer candidates from the dance party could be found. It could not. The *Scarlet and Black* did not report it, nor was I shown anything similar to the minutes of Talladega's Prudential Committee that might have recorded it. But I remember it. The entire episode—the exclusion, the rationalization, the dismissal from mind and the apparent omission from the record—has shown me how pervasive the force of racial prejudice could be, reaching into a college founded by abolitionists, and what a respectable face it could present to the world.

Georgia Patterson believed she had encountered that same

force in her college. For speaking her mind about it—vehemently, it is true—she was deemed unworthy of remaining in school. The complaints she voiced in her first interview with President De Forest sealed her fate, as far as he was concerned. And my grandmother, alas, gave her consent to Georgia's removal from school without hesitation.

Amorette would live fourteen more years after those turbulent days in March of 1895. Did she ever remember Georgia Patterson, as I still remember the Negro boy in my Latin class? Did she ever wish—regretfully, as I wish—that she had not done what she did? The question is rhetorical, of course. I do not know what thoughts came to Amorette when she was old and had time to ponder actions she had taken in the past. But I know she worked for an organization that denounced racial prejudice and she believed in a religion that preached charity and forgiveness. The writings she left behind—they will appear in a later chapter—tell me that she had a tender heart. For these reasons I will let my rhetorical question stand.

Chapter 23
Departures

The protest and its aftermath had consumed the attention of the Prudential Committee for ten days, but at the March 9 meeting, after they heard President De Forest's report, they turned to other matters. The repairs to college buildings, which a year earlier had been called "of first importance and necessity," had still not been made. Now the Committee was almost ready to send the cost estimates to the American Missionary Association for approval. One more thing remained: to divide the list into items of "absolute necessity" and those of "secondary importance." I hope that some of Amorette's requests for Foster Hall—the new roof, at least—made the first list.

That spring Dell, acting on his mother's behalf, reopened the long-dormant question of Seth's muster into the army. He had learned of an act of Congress that offered "certain officers of the army" the hope of an adjustment in the date of their promotion to a higher rank, which in turn might lead to an increase in the amount of their pension or that of their widows. Encouraged by a Washington lawyer, Dell filed a claim for Amorette. He signed it in her name—I recognized his handwriting—and gave his address in Omaha as hers. In authorizing Dell to make this claim, Amorette was probably still clinging to the mistaken idea that Seth was mustered in as second lieutenant when he enlisted in the army in August 1861. Dell knew the facts, however, and correctly stated that Seth was promoted to that rank in December. The Pension Office, citing the December muster, denied the claim. Amorette was disappointed; Dell, I should think, was embarrassed; and her pension remained at fifteen dollars a month for the rest of her life.

The only good thing to come out of this episode—and it was a long time coming—was that it told me the address in Omaha where Dell and his family lived and where Amorette was spending her summers. Looking up 1130 South 31st Street on a city map, I saw that it was very close to Hanscom Park. I remembered that park well because my best friend, Marjorie Moore, lived across the street from it. One winter day in about 1939 we tried to go skiing there. She had borrowed, with or without permission, a pair of skis that belonged to her older brother Dinty. We carried them across the street and into the park, where we found a gentle incline that looked suitable. Neither of us knew anything about skiing.

It was decided that I would take the first turn with the skis, so I fastened them on and stood up. Summoning the courage to move my feet, I began to slide slowly, inexorably toward a little hollow, at the bottom of which stood a tree. Naturally I did not know how to turn away from it. But my balance was so unsteady that before the tips of Dinty's skis could run into the tree trunk I fell over and sank into the snow, where I lay helpless with laughter until Marjorie, almost as helpless, pulled me to my feet.

I spent a good deal of time in the Hanscom Park neighborhood during those years. Marjorie and I would ride to her house from high school on the streetcar, which ran along the edge of the park and stopped at her corner. She had a big house and a fairly big family. When I sat down to dinner at the Moores' there were seven of us. My household, by contrast, consisted of only two members, for my father had died several years earlier. I loved being part of the Moores' family, and I did not consider, I'm afraid, how many lonely evenings my mother must have spent. In the morning Marjorie and I would take the streetcar back to school, carrying the lunches the maid had packed for us. (At home I packed my own lunch.) And in all the days and nights I happily spent at Marjorie's house I had no idea how close I was to the place where my father and his other family and his mother had once lived.

In the fall of 1895 Amorette began her fourth year as matron of Foster Hall. By November she had succeeded in clarifying the question of who should or should not eat in the Foster Hall dining room. There had been confusion in the past over this question: a male student, apparently dissatisfied with the food, had asked the Prudential Committee if he could eat his meals "elsewhere than in Foster Hall dining room." Yes, responded the Committee, if he did not room in the college. Now, two years later, a new rule made the policy official: a student who does not eat in the Foster Hall dining room may not room in a college residence hall; conversely, a student who does not room in a college residence hall may not eat in the Foster Hall dining room. From Amorette's standpoint the new rule, which allowed for a more orderly allotment of money and procurement of food, would make the operation of the dining room and kitchen more efficient. From the students' standpoint, of course, it would constitute one more limitation of their personal freedom.

The college suffered a terrible blow in late January, with the sudden death of President De Forest. On Friday, January 24, as the local newspaper recounted, he had seemed to be in good health, for he was seen walking about town. An early riser, he got out of bed at five o'clock the next morning and kindled a fire. Then, just as he reached the head of a flight of stairs, he had a stroke and fell down all twenty steps, to lie unconscious at the bottom. He lived two more days, never regaining consciousness. The funeral service, conducted by Dr. Andrews, took place on the following Tuesday. *Our Mountain Home* reported that one of the pastors who assisted was the Reverend Mr. Brown ("colored"), a graduate of Talladega College. Of the black population of the town, the paper wrote, "Many are they who are saddened over the sudden death of their benefactor." De Forest had guided the college with a firm hand, and if it occasionally had seemed too firm, that quality was forgotten in the outpouring of communal grief.

Still reeling under their loss, the Prudential Committee nevertheless had to go on enforcing the rules. At a February meeting it voted to reprimand Daisy Siegel "for breaking Mrs. Foot's Dining Room rules." Daisy was one of the nineteen girls who, considered to be lesser offenders, had been "reprimanded, admonished and restored to their position" in school after the protest of the previous years. Now she was to be reprimanded but no further punishment was imposed.

Amorette's second granddaughter was born in late March and was named Mildred. Amorette spent the summer of 1896 as usual with Dell and his growing family; then in September she stepped off the train at the Talladega depot and back into her institutional life. The Prudential Committee undoubtedly went on meeting that year and the next two, and Amorette almost certainly continued to serve on it, but I cannot report on its actions because the minutes for those years could not be found when I visited the Talladega College Archives. What I did find, covering my grandmother's last three years at Talladega, was a small collection of news items.

In January of 1897 *Our Mountain Home* reported that a "very enjoyable" concert was given at the college and that "the attendance was good." This item, like the respectful obituary for President De Forest, suggests that the antagonism toward the college, once felt by the white people of the town, was gradually being replaced by acceptance and civic pride. Later that year two separate fires in Foster Hall gave Amorette cause for anxiety. The first one was the more serious; it destroyed the laundry. It is difficult to imagine how and where the washing was done for the next few years, since a new laundry was not built until after Amorette left. The other fire was much smaller and is noteworthy mainly for the quick action of one of the young residents of Foster Hall. On a December night Willie Elston, "a Grammar girl" (meaning that she was taking the elementary course), was studying in her room when her lamp suddenly exploded. Thinking fast, Willie put out the flames before they spread.

Amorette gave up her job in June of 1899, after seven years of service. In this college, where turnover was high, just staying that long was an accomplishment in itself. During her tenure three different women had held the position of matron of Stone Hall and four had held the position of preceptress. At the time she left she was in her sixtieth year, and she was feeling her age. *The Message and Deaconess Advocate* reported that "Mrs. A. E. Foote, tired with her many years of matron work, is resting at Clifton Springs, New York." Amorette's choice of destination shows that she was still concerned about her health, which probably had not benefited from her stay in the South after all.

Clifton Springs Sanitarium, located in a valley watered by sulphur springs, offered a regimen of warm baths, Turkish baths, salt baths and massage, combined with healthful diet, exercise and pleasant recreation. Upon her arrival Amorette would have seen a five-story building, like Foster Hall constructed of red brick, but considerably grander. It boasted not only a solarium on the top story but electric lights, comforts she could scarcely have imagined during her years at Talladega College. Even the grounds seemed designed to bring health to body and spirit. Walking paths, leading to pools and ponds, stretched across green lawns; under the shade of a small pavilion, benches faced each other to encourage conversation.

Amorette could have taken comfort too from the Methodist influence she could sense around her. Though the Sanitarium was not a religious institution, the owner, a Methodist, had built a chapel on the grounds. The presence of a Methodist church in the village would also have helped to make her feel at home in Clifton Springs.

There was a stronger force, however: here in this valley in Central New York she was close to the place of her birth, closer than she had been since she left it as a four-year-old child. For Lenox Township was only seventy-five or eighty miles east of where she was now, and the old Erie Canal, a fixture of her early childhood, lay just five miles north of Clifton Springs. She had come back to a land she remembered perhaps dimly, but being there might awaken more memories. When I think of my grandmother resting at Clifton Springs, I do not visualize her steaming in a Turkish bath or lying on a table while a masseuse kneads her tired muscles. Instead I see her sitting, fully dressed, on a bench in a small pavilion, remembering her first home.

Southern California Conference Deaconness Home
Courtesy General Commission on Archives and History,
The United Methodist Church, Madison, New Jersey

Chapter 24

Amorette's Voice

When she took on her next job, Amorette re-entered the world she had left when she left Chicago—the world of the deaconesses, a world of women pledged to care for the needy and spread the gospel of Christ. In the summer of 1901, after serving as superintendent of the Southern California Deaconess Home in Los Angeles for more than a year, she submitted the first of her annual reports to the Woman's Home Missionary Society of the Methodist Church. In this report and those that follow year by year, I finally hear a voice that is clearly Amorette's. Up until now, the statements that have come from her, with the exception of the one word "Yes," have been mediated by someone else. A lawyer wrote her depositions for the Pension Office, an editor wrote her thumbnail biography. But her job in Los Angeles required her to write her own account of what her deaconesses had been doing, and she complied fully and enthusiastically:

"Southern California Deaconess Home
180 Hewitt Street, Los Angeles, California
Mrs. A. E. Foote, Superintendent

During the past year we have realized many blessings, and have tried to pass them on to others; so, by having our lives become richer and sweeter with God's manifold blessings, we can uplift and help those with whom we come in contact to be more like him. We have six deaconesses in our Home besides the two who are connected with it. Our nurse deaconess, Miss Hoover, who came to us last fall, has given excellent service to the sick ones, and with her cheery face has relieved many an aching heart as well as body."

In what was probably Amorette's first experience of writing for publication, she employs the fervent tone I have heard in the writing of Lucy Rider Meyer and others associated with the Chicago Training School. Amorette does not seem entirely comfortable with it, yet her sincerity cannot be doubted. In her second paragraph I hear her admiration for her deaconesses and her sympathy for those to whom they minister. But her reliance on sentimental clichés causes me to wonder if this was the way she talked:

"Mrs. Parcell has found her field of labor in Fresno, where are many needy ones to whom she ministers and gives the enlightment of the gospel. Miss Wilbur and Miss Chaffee have been assisting the pastors of the city with the visiting; they also give help to strangers, the sick and needy ones, and bring the children—'God's little ones'—into the Sunday-school, sewing-schools, and Junior League. Miss Bisbee, our Traveler's Aid, was up at the break of day to 'be about her Father's business,' in helping any needy traveler. A record of the lives that were saved from ruin, and the lonely hearts cheered, will not be forgotten when the books are opened on that great day. Miss Phillips, who has been going out in the byways helping the pastors to garner in the sheaves, has had many opportunities to speak a word of help to those who were hungering and thirsting after righteousness, and through her many have found the satisfying portion."

Here Amorette provides an informative survey of the tasks the Los Angeles deaconesses have carried out during the past year. I can't help remarking, however, that she takes her own missionary role over-seriously. In her last three sentences she has managed to draw upon four different books of the Bible—first Luke, then Revelation, then Psalms and finally Matthew. This seems excessive, even in a Christian organization. Yet in the very prolixity of her initial report I hear the voice of a woman doing her best to put into words the soul of the task she and her sister deaconesses have undertaken.

At the Los Angeles Home she could wear her deaconess bonnet again, with pride. I doubt if she wore it while she was at Talladega. If she had, she probably would have been the only person in the college to do so, and the symbolism of it would have been lost on faculty and students alike. But all the residents of the Hewitt Street house were entitled to wear the small black bonnet with the white ties, and here again, as in Chicago, it gave the wearer protection when she ventured into unknown, perhaps intimidating neighborhoods. Amorette may not have worn the bonnet when she was inside the Home, but she surely would have put it on to go outdoors, and perhaps she wore it indoors when officials of the church came to visit.

The Deaconess Home, less than a year old when Amorette arrived, replaced a succession of rented properties which the Los Angeles deaconesses had used during the first four years of the group's existence. The lot on which it was built, valued at one thousand dollars, had been donated by Mrs. W. C. DePauw. She was the widow of Washington C. DePauw, a wealthy manufacturer and devout Methodist, whose substantial donations to a Methodist college in his home state of Indiana resulted in its rebirth as DePauw University. After his death Mrs. DePauw moved to Los Angeles, where she continued her husband's practice of giving to Methodist causes.

The lot that Mrs. DePauw chose was located about two and a half blocks from the Santa Fe railroad tracks. At the time the Woman's Home Missionary Society received the gift, the railroad station was the most famous landmark in the neighborhood. La Grande Depot, known also as La Grande Station, was a rather astonishing red structure of Moorish design, topped by a bulbous golden dome. Since its opening the depot had brought crowds of arriving and departing passengers, as well as a steadily increasing freight traffic, to the neighborhood. Warehouses and other commercial buildings lined the streets between the station and the deaconesses' lot.

The building of the new Home was accomplished with surprising speed in the first half of 1899. Branches of the W.H.M.S. from all over southern California had been giving or pledging sums of money. Mrs. DePauw came forward with another thousand dollars, though this time it was a loan, with the threat of a ten percent interest charge if it were not repaid within the stipulated time. In early July the new Deaconess Home opened.

It was a handsome house, square, two stories high. In a photograph taken shortly after its completion it has a look of solidity and importance. A flight of steps, narrowing as it ascends, leads to a broad front porch. Above the entrance a triangular pediment announces DEACONESS HOME in capital letters that follow the line of the porch roof, the apex of the triangle rising between the N and the E. The exterior surface of the house is covered with clapboard siding, painted in a light color; a darker color outlines the eaves, the window frames and the pediment. Set down uncompromisingly on its level plot, the new Deaconess Home still looks somewhat raw; no shrubs soften the line where the foundation meets the ground. But grass is already growing in the spot of front yard that is visible, and someone has placed a rocking chair and a straight chair on the front porch, for those rare moments of leisure at the end of a work day.

I said earlier that in taking the job in Los Angeles Amorette was re-entering a world she had been in before. The statement, while true, needs some qualification. Amorette would not be working in the deaconess organization founded in Chicago by Lucy Rider Meyer. The Woman's Home Missionary Society, also connected with the Methodist Church, was founded by a former college professor, Jane Bancroft Robinson. Unfortunately a rivalry soon developed between the two organizations and their leaders. It was to last for many years, provoking strong feelings on both sides.

I do not think, however, that Amorette chose the W.H.M.S. for political reasons. I think it is more likely that her motives were personal and geographical. Southern California at that time enjoyed a nation-wide reputation as a center for the restoration of health. Amorette's concern for her health has been evident from the time she took the job in Alabama. I think that she, encouraged probably by Dell, believed California's well known sunshine would surely be good for her.

In the fall of 1901 the Deaconess Home had a visitor, a missionary who was returning to Africa to resume the work she had been doing for the past fifteen years. This was Susan Collins, the African-American woman whose footsteps Amorette had unwittingly followed from Fayette to Dakota to the Chicago Training School. The occasion was the annual meeting of the Pacific Branch of the Woman's Foreign Missionary Society, the Methodist group that was sponsoring Miss Collins's return to Angola. A Los Angeles woman who had met Collins years before in Chicago wrote an account of the visit for the *California Christian Advocate:* "I had gone to our pleasant Deaconess Home to see her off, but she was too busy with friends to spend time to sit at the table with us. The Superintendent could only put up a generous lunch for the quiet hour after the train was under way."

I envision Amorette in the dining room of the Deaconess Home. Seeing that Miss Collins has not touched the food on her plate, she hurries back to the kitchen and puts together a substantial lunch (for the train trip will be long and meal service uncertain). She brings it out and offers it to Miss Collins, who takes it and thanks her. The two women shake hands, and then it is time for the visitor to leave for the train station. Did they ever have the opportunity to talk about Fayette? Perhaps. Did they happen to discover that both of them had been born on the day before the Fourth of July? Susan, born in 1851, in a free state, to an escaped slave; Amorette, born in 1839, widowed by the war that freed all the slaves. . . Probably there had been no time for such reflections.

Writing her annual report for 1901-02, Amorette seems happy with the atmosphere in the Home and optimistic about its future: "The past year has witnessed large growth in our work, due, under God's blessing, to the wise management of our efficient Board and to the earnest labors of our deaconesses. Our Home is on a good financial basis, being free from debt, and is now in a position to reach out toward an enlargement of its borders. Thus it has been decided to purchase the property adjoining the Home, to be used as a Deaconess Hospital. Peace and harmony have prevailed in our Home, and it is with pleasure that we note greater interest in the work, not only by the Churches, who are awakening to the need of deaconess work, but also by the people in general."

In the year since she wrote her first report Amorette has learned to express herself with greater economy and restraint. She still acknowledges God's presence in their home but she does not labor the point. She presents financial information persuasively, making clear, to insiders at least, that the debt to Mrs. DePauw has been paid. Her statement about peace and harmony suggests the possibility that disharmony might arise in a home like this one, if it were not met with "wise management." Behind her phrase, I believe, is the assumption that she recognized wise management when she saw it because she knew how to provide it herself.

As she continues, she touches upon the organization's policy regarding marriage for deaconesses: "We have had seven deaconesses with us during the year, with two still out on leave of absence. Two of our members have found another field of usefulness in homes of their own, two new ones have come to us from other fields, and we anticipate the coming of our nurse deaconess, Miss Hamilton." Amorette's phrase "in homes of their own" reminds me of the words Isabella Thoburn wrote years before. Thoburn said "the happiest place" for a woman was a home of her own with husband and children; the deaconess home was only "the next happiest place." But Amorette will not allow any ranking. To her the two kinds of life are equal in value, marriage being simply "another field of usefulness."

This year Amorette's closing prayer enters her discourse

more naturally than in her earlier report: "We are truly thankful to our Heavenly Father for his many blessings, and we earnestly hope and pray that during the coming year many more needy ones may receive both temporal and spiritual aid." Mercifully, she does not find it necessary to pull in so many biblical allusions. In fact there are none in this report, so that her own voice, devout but not ostentatiously so, carries a more personal conviction. I cannot entirely believe in the work the deaconesses were doing; specifically, I can't approve of their efforts to draw their clients away from their own religion. But I am convinced that in their humanitarian work and their ministry to those in need of a faith, the deaconesses brought aid and comfort to many people.

Amorette, in turn, saw it as her duty to provide aid and comfort for her deaconesses. Evidently she succeeded in her task. In April of 1903 one of the Los Angeles deaconesses paid public tribute to the spirit Amorette brought into their home. Writing for the *California Christian Advocate,* she says, "Mrs. Foote, our able Superintendent, does not lack for opportunities to use her willing heart and hands. She is to this household all the word 'mother' implies, and how a little 'mothering' does help to lighten our burdens."

The writer has seen in Amorette the trait that I, tracing her life story, have been observing. For nearly thirty years her maternal love flowed mainly toward her son, though perhaps her student boarders in Fayette felt some of it. When Dell left Chicago and she stayed, she became an institutional mother, a role she continued to play in Alabama and California. But *role* is surely the wrong word. She was not playing a role, she was living the life that had been given to her, the life she felt was naturally and comfortably her own. Today a woman like the writer for the *Advocate* would hesitate to use the concept of "mothering" as a term of praise, either because it sounded too sentimental or because its use in this context would tend to diminish her own independence and that of her fellow deaconesses. But times were different then. The early 1900s were still close enough to the nineteenth century that the idealized figure of the mother, memorialized by Catharine Beecher among many others, commanded an almost automatic respect. The woman who wrote this tribute to Amorette had a clear image of all the word *mother* implied—all the skills, all the virtues—and she could assume readers would recognize it and share her reverence for it.

The annual report for the 1902-03 year was written by two persons, Amorette and one of her deaconesses. In the first paragraph Amorette speaks in her familiar voice: "The past year has abounded in good works, though we have been unable to respond to the many urgent demands made upon us because of the limited supply of workers. Our five deaconesses are not required to do the work of ten, yet when I listen to their reports, I am led to think that they are trying to do so."

The first two sentences of the next paragraph still sound like Amorette: "All of our undertakings this year have in a measure been successful. God has blessed us abundantly." But the next two are obviously spoken by someone else: "With one exception our deaconesses have closed the year in better health than they began it. It is with the deepest regret that we are obliged to grant our beloved Superintendent a year's leave of absence on account of broken health."

Broken health! The alarming statement made me think that Amorette began her leave in 1903, but other evidence shows that she remained on the job for nearly another year. An article in the *Deaconess Advocate* for April 1904 places her in the Hewitt Street house that spring: "Los Angeles.—The English work carried on by eleven deaconesses, comfortably homed in the commodious house on Hewitt street, is very quiet, but very effective. Mrs. A. E. Foote is the efficient superintendent." The writer sounds like Lucy Rider Meyer, paying a visit to her alumnae: "There are familiar faces among these deaconesses, Mrs. Foote, Miss Decker, Miss Steele and Miss Lowrey, all from the Chicago Training School." Surely Mrs. Foote's "familiar face" would not have been included unless Amorette had been present.

In the summer of 1904 Amorette wrote a full and knowledgeable report for the year past. Her name appears in the heading, and the voice is unmistakably hers—sympathetic and personal, a bit sentimental: "Our Home in Los Angeles is in a flourishing condition. The people of the Churches and many others are interested and helpful. In our hospital, which is just in the bud, only opened last October, we have more patients sent to us than we can take in for lack of room. Our two nurses are ever busy, as they attend patients out in the district as well as those within our doors."

This report and subsequent ones use a new address: 226 South Hewitt Street, rather than 180 Hewitt Street. Perhaps the city of Los Angeles altered its numbering system; certainly the deaconesses did not not move out of their first house and into another down the street. Taking the 226 number as the corrected one has enabled me to place the Home on the map, with a fair degree of certainty, between East Second and East Third streets. Half a block north of the Home, Second Street led directly to the main entrance of La Grande Depot.

"Two of our deaconesses work at the Santa Fe and Southern Pacific depots as travelers' aids," Amorette continues, "and many are the sick and forlorn whom they are able to help. The

Young Women's Christian Association partly support these workers and have the supervision of their work. A club of self-supporting girls of this Association pay the rent of a Rest-room for the use of the travelers' aid deaconesses in their work, and it is there many a weary soul rests after a long journey, or waits for belated trains, and often, if out of money, remains in the room, rent free, until work can be procured for her." This provision of shelter and assistance for penniless female travelers bespeaks an uncritical generosity that would rarely be matched today and may have been rare then.

As she ends her report, Amorette calls attention to the role of the deaconesses as teachers of English. "Our visiting deaconess, who works among the Japanese, is teaching the women in their homes the English language and finds much encouragement." In going about "the Japanese work," "the Chinese work" and "the Spanish work," the W.H.M.S. engaged teachers who were fluent in these languages. Predictably, the instruction combined English lessons with efforts to convert the pupils to Protestant Christianity. They were least successful among the Spanish speakers, who it was said appreciated the vocational schooling and the music but not the proselytizing. A leader in the Spanish work laid the blame upon "the strong Catholic influence."

The year 1904-05 must have been the one in which Amorette began an extended leave of absence. In late December 1904 a Mrs. M. R. Spence assumed the position of superintendent of the Los Angeles Conference Deaconess Home, but she held it for only a short time. When the report for 1905-06 was written, "Miss Birdsall" signed it as superintendent. Seeing her name, I remembered that M. Jennie Birdsall was a student at the Chicago Training School while Amorette was matron there. Now Miss Birdsall writes a brisk report, filled with facts and figures; "fifteen deaconesses. . . have made 10,844 calls, . . . ninety patients have been cared for in the hospital and thirty-six in homes." Though reserved and business-like in manner, she is obviously gratified to note that "our Japanese worker is doing much good among the the Japanese women, teaching them English and leading many into true and saving faith."

Absent from the record for the previous year, Amorette has now returned, but to a different place. She writes as superintendent of a newly opened "rest home" for deaconesses, located in Huntington Beach: "The Wing Deaconess Rest Home is situated on a fine lot about four blocks from the ocean. It is a six-room, modern cottage, is nicely furnished throughout, and is free from debt."

In the parlance of the deaconess organization the term "rest home" did not mean what it commonly means today. It is true that the Methodist leadership was beginnng to feel concern over the impending problem of "superannuated deaconesses," but the Wing home was not intended for them. This seaside retreat, situated on the Methodist camp grounds thirty miles south of Los Angeles, was designed as a place where a working deaconess could go for a brief period of rest and refreshment of spirit. A benefactor identified in the records as "Brother Wing" gave the W.H.M.S. a large share of the money for its construction. At the time Amorette wrote her report, the Wing home had been in operation about three months and seemed to be serving its purpose admirably: "This rest home has brought renewed strength to many tired out workers since its operation. Eleven deaconesses have been entertained in the Home. The purpose in building this Home was to have a quiet, restful place which would be absolutely free to the deaconesses under our care." So far Amorette has said nothing exceptionable, but in the next sentence she made a statement that would get her into trouble: "All workers outside the Conference pay a nominal price when entertained in the Rest Home." In allowing, even inviting, workers outside the Southern California Conference to stay in her cottage, Amorette was evidently extending a privilege too widely. Repercussions from her well-intentioned statement would be felt for the next two years.

The last piece of her writing that I have is my favorite. It is not an official report, just a personal note in the *Deaconess Advocate,* published in June 1907, when she'd been superintendent of the Wing home for a little over a year: "My work is caring for our little rest home and mothering the girls when they are tired out and come to rest. The Home is a six room cottage. I have a garden, seven hens and eleven little chickens. I believe I am the oldest deaconess in the work—will soon celebrate my sixty-eighth birthday." In this passage it is clear that she is thinking of "the girls" as her daughters. Addressing her old friends at the Chicago Training School, she speaks for herself and about herself. This is how Amorette sounds when she looks at her life, contentedly, and tells her readers what she sees. Her words give no indication that she is about to leave this job and I don't believe that was her intention.

I treasure these four sentences of hers, even as I am gently, indulgently amused by them. Here is my grandmother, living only four blocks from the Pacific Ocean, and she seems to look no further than her own garden and her own chicken yard. She might as well be back in Fayette. . . . I remember when I lived four blocks from the Pacific—I went down to the ocean every day. I was taking my small son to the beach. A picture of us forms in my mind: a little naked boy, not yet two years old, running full tilt into the surf; a young woman, her long hair flying, scrambling across the wet sand to snatch him up before the next wave knocks him over. . . .

Well of course Amorette couldn't go to the ocean every day. She was almost sixty-eight years old, and she had her own kind of mothering to do. But by the time her birthday came around she was no longer entitled to do it. At the end of the year 1906-07 the superintendent's annual report on the Wing Rest Home was not written by Mrs. A. E. Foote. That duty was performed by Miss Birdsall, who in her first sentence corrects the statement Amorette had made the previous year. "This Home," Miss Birdsall writes, "was provided for the use of the deaconesses of the Los Angeles Deaconess Home only, and is under the same management." In other words, she and not Amorette is in charge of the Wing Rest Home, where no outsiders are allowed. It must be true, then, that shortly after Amorette's note was published she left her position at the Wing Rest Home and withdrew from active deaconess work. I suspect that someone in authority had decided sixty-eight was a suitable age for retirement.

The next year the Wing Rest Home had a new superintendent, Miss M. Eleanor Rich. She makes the correctlon even more emphatically than Miss Birdsall had done: "The Home is for deaconesses in the Southern California Conference only." But even though Miss Rich went out of her way to correct her aunt's mistake (for she is the daughter of Amorette's brother Angello), she has adopted Amorette's fervent tone: "It has been a haven of rest to many a tired worker," she writes, "who thank God for its protecting care." Presumably Miss Rich, known in my father's family as Cousin Ella, had been inspired to join the deaconess service through Amorette's example.

By the summer of 1907 the Los Angeles deaconesses had moved out of the house on Hewitt Street. They were living temporarily in a rented place on South Westlake Avenue (more residential, less industrial), and they planned to buy a permanent home "in a more desirable section of the city" as soon as the Hewitt Street house could be "disposed of to good advantage." It is sad to think that the home on which so much effort and money had been expended, and in which so much love had been given and received, was to be disposed of so soon.

At the time, however, the reasons for the move were compelling. They had been spelled out to the Conference management in 1906 by Will Knighten, a member of the Deaconess Board, who complained of the discomfort and inconvenience of the Home's location. "The noise, dust and general confusion," he wrote, "renders it almost uninhabitable for women, and the health of our Deaconesses rapidly fails when residing here any length of time." These conditions may have contributed to Amorette's "broken health" a few years earlier, necessitating her leave of absence and prompting her transfer to an easier job in a quieter place.

The location was ill-advised from the start. Though Mrs. DePauw's lot had cost the deaconesses nothing, it turned out to be a bad bargain. But for a few short years the Hewitt Street house was the dwelling place of kindness and concern for the unfortunate and perseverance and, yes, zeal. At its center stood Amorette, holding things together.

Chapter 25
Her Homecoming

For some time Amorette had been considering Omaha as her home, even though she spent most of every year somewhere else. But from now on, that city, and her son's house in that city, would be the only home she had. When she moved in permanently with Dell and his family, they were still living in the house on Thirty-first Street. Amorette's arrival may have crowded them somewhat. It was one thing to make room for a visiting grandmother who was sure to leave at the end of the summer and another to realize that she was there to stay.

I will assume that one bedroom in the house had always been reserved for Amorette, since her visits over the years had been so frequent. In this room, I think, is where she would have had her rocking chair. It is of the right age to have been new in 1907; an antique dealer to whom I once showed it told me it had been made—factory-made, he said—early in the twentieth century. I envision Dell buying the sturdy, comforting rocker for his mother when she moved in. Having placed this real chair in this imagined bedroom, I can now resurrect the image of my father's mother that I formed in childhood: an old lady with a gold watch in her pocket, sitting quietly in her rocking chair. But probably not wearing her bonnet, as I thought when I was young, and certainly not consigned to an institution, as I also thought.

Dell, now in his late forties, was leading an extremely busy professional life, seeing patients daily in his office, performing surgeries at two Omaha hospitals, delivering babies at all hours of the day or night. In the evenings he made house calls, driving his new automobile. Still he found time to serve as president of a national obstetrical society and the alumni association of the Chicago Homoeopathic Medical College. Of necessity, the home life of this family must have revolved around the father's needs and wishes.

About a year after Amorette's arrival the family moved to another house, still in the Hanscom Park neighborhood, at 1313 Park Avenue. Perhaps the new house was larger, chosen in the hope that greater space and more privacy would reduce the friction between family members. For there was friction—in particular, friction between Dell and Milla—or so my mother said. Relying on what my father told her, years later, Mother blamed Milla for provoking the quarrels. I can have no opinion on the question, but as I peer into the home that I have reconstructed, I see elements that could have caused tension and ill-feeling. The bond of love and loyalty between Dell and his mother was very strong and of long standing. Once the two women were permanently confined to the same house, differences would almost certainly arise. Amorette, having managed households much larger than this one, may have found it difficult to give up her authority. Dell may have sympathized with his mother; Milla may have felt that she did not come first in her husband's affections. In times of disagreement both Dell and Milla might have remembered the misgivings each of them had entertained before their marriage. The situation in which these three people had placed themselves invited dissension, it seems to me. I know that taking a mother/mother-in-law into a family's home was much more common a century ago than it is today, but I cannot believe that the atmosphere in such homes was always peaceable.

Though retired, Amorette did not relinquish her identity as a deaconess. Many years earlier her description of the organization, conveyed to the Omaha Methodists by Dell, had played a part in the formation of the Omaha deaconess program; now, in her old age, she took pleasure in following its progress. On a night in late May, 1909, she attended the graduation ceremony for nurse-deaconesses at the Omaha Deaconess Hospital. She enjoyed it "greatly," it was said.

On her return she walked into the house, and before she even had time to take off her deaconess bonnet she suffered a stroke and fell to the floor unconscious. She lived a few days longer, never regaining consciousness. On June 2 she died, one month and one day short of her seventieth birthday. A skull fracture, incurred in her fall, contributed to her death, but it was the stroke that killed her.

Dell sent the news of his mother's death to the Chicago Training School, where someone, probably Lucy Rider Meyer, wrote a notice for the *Deaconess Advocate.* It ends with the words, "She rests from her labor, 'and her works do follow her.'" This simple paraphrase of a verse from the book of Revelation was the most fitting epitaph that could have been written for Amorette.

Epilogue

On a day in early June they take Amorette's body to Walnut Hill Cemetery in Council Bluffs. The hearse travels over the Missouri, then east across a level plain toward the wooded hills. It is a place much like the one where Seth and the rest of the company had camped nearly fifty years before: same river, same cottonwoods and oaks, same bottomland extending several miles to the base of the bluffs. The goldseekers had taken a day of rest when they reached the Missouri Bottoms, even though it was not the sabbath. They strolled along the river; they admired the grass and the trees and the rich fertile soil; they did some desultory hunting. Now, as Amorette goes to her rest on a green hillside, there is no one among the mourners who remembers that earlier day, no one to invoke Seth's spirit.

The hearse toils up a narrow winding road to the Rich family plot, where Amorette's brother Newton was buried two years earlier. There is still plenty of room for the rest of Newton's family, when the time comes. Today there is room for Amorette, who has died so unexpectedly.

The ready availability of the space was probably what led Dell to bury his mother here. What it would come to mean, however, was that on every Memorial Day he would have to drive eight or ten miles—no small undertaking in those days—to lay flowers on his mother's grave. Still he kept up the custom as long as he was able. On May 30, 1923, he wrote in his diary, "Went to Walnut Hill Cemetery Co. Bluffs early & decorated Mother's grave—also Uncle Newton's Aunt Susan's & Paul's" (Paul being the son-in-law of Newton and Susan). The matter-of-fact tone of my father's entry suggests to me that the Memorial Day ritual was an annual event, but I do not mean to imply that his yearly decoration of his mother's grave was perfunctory. His regard, amounting to reverence, for her memory makes itself felt in his entry for July 3: "Mother's birthday. She would have been 84 years on earth & now has been 14 years in Heaven."

He wrote these words in the first year of his marriage to my mother. He had known her, I think, for about ten years, ever since she was a student nurse in a hospital where he practiced. Some time during those years he left his wife and family and went to live in a hotel close to his office; Milla moved to California; he and Milla were divorced. I do not know in what order these events took place, only that Mother, for most of her long life, almost never talked to me about them. But once when she was quite old she told me emphatically that he had moved out of the family home before. . . What she said next is not clear to me and wasn't then. I think she meant before all the rest of it happened. Before he and she publicly acknowledged their love for each other. Long before the divorce, which took years to accomplish.

It was a great love, though outwardly incongruous, this love between the young nurse and the middle-aged doctor. In the letter that accompanied her engagement ring Dell wrote, "I ask God tonight to bless our hearts & give to our love the spirit of constancy that will disprove all statements of critics or scoffers. . . The past is gone. The present is full of heartfelt & laborious duties. The future is radiant with love. God is over all. We bow our hearts to him & humbly wait."

Dell's profound faith in God must have been instilled in him by Amorette. It was one of the ways in which she carried out the mandate—"Raise virtuous sons!"—that was laid upon American mothers of her century. Yet in the years following his mother's death Dell committed a grave offense against the principles of his religion. Jesus taught that a man who divorces his wife and marries another woman is guilty of the sin of adultery. "Whosoever shall put away his wife and marry another," he preached, "committeth adultery against her" (Mark 10:11). Amorette, steeped in the New Testament, would have known these words and subscribed to them. I think she would have been greatly saddened by Dell's disregard for Christ's teaching. Mercifully, she died before she had to know of it. I of course cannot be sorry that my father did what he did. If he had not, I—the exact person that I am—would not have come into existence. I am the granddaughter my grandmother never knew, and that fact too was merciful. My life, with its record of marriages and divorces, would have shocked and baffled her. But I think her loving nature would have impelled her to forgive us both. Dell first, because he was her son; but Virginia too, as one of the "little girls" who came under her care.

I only think, I do not know. What I do know is that in

bringing my grandmother out from the shadows into which she had receded I have come to love her. Someone may object that the Amorette I love is a woman of my own creation, but I maintain that she was there all the time.

Comments on Sources

It must be obvious to readers that many sources went into the writing of this book. They vary widely in nature: two diaries; some old letters; census records; archival materials housed in California, Iowa, South Dakota and Washington, D.C.; small-town newspapers; the reports and newsletters of religious organizations; books, with publication dates ranging from the 1840s to the twenty-first century. In this essay I will identify as many of them as space permits. To avoid repetition I have tried to list each source only once, even when it was used in more than one part of the book.

Part One: The Early Years

There exists, in usable form, a record of vital statistics for each of the two families who figure prominently in Part One. For the Foote family it is a 600-page book by Abram Foote: *Foote Family: Comprising the Genealogy and History of Nathaniel Foote of Wethersfield, Conn. and His Descendants,* Vol. 1 (1907). For the Rich family it is a single page, entitled "Family Record," listing births, deaths and marriages from the birth of Amorette's father in 1799 to the death of her husband in 1864. It is contained in the "Pension Application File for the Widow of Seth Foot" in the National Archives in Washington, D. C..

County deed indexes record purchases of land, and census forms yield information about occupations and economic status, among other things. A biographical sketch in an anonymous *History of Fayette County* (1878) summarizes the moves

Amorette made from 1844 to 1861, when she came to Fayette with her mother and son. Moses Foot's obituary in the *Cherokee Weekly Times* (1894) provides facts about his early life.

Material for the setting of Amorette's first home is drawn from Laurence Hauptmann's *Iroquois Dispossession and the Rise of New York State.* Facts about boat travel come from *Erie Water West: A History of the Erie Canal 1792-1854* by Ronald E. Shaw and from *American Ferryboats* by John Perry. Francis P. Weisenburger's history, *The Passing of the Frontier: 1825-1850,* describes life in Ohio during the time the Rich and Foote families lived there.

Two books on Iowa history contribute information relevant to the families' emigration to Iowa: *A History of Iowa* by Leland L. Sage (1987) and *Iowa on the Eve of the Civil War* by Morton M. Rosenberg (1972). N. Howe Parker's Iowa *As It Is in 1855* exemplifies the hortatory guidebooks that were offered to prospective emigrants. The debts incurred by Seth Foot and Philander Austin are documented in Seth Foot, Will and Supporting Documents, held at the County Recorder's Office, Clayton County Courthouse, Elkader, Iowa. *The Emergence of Lincoln* by Allan Nevins attests to the severity of the depression in the Upper Mississippi Valley in the late 1850s. *"This State of Wonders": The Letters of an Iowa Frontier Family, 1858-1861,* edited by John Kent Folmar, illustrates its discouraging effect on individual farmers. The anonymous author of *History of Clayton County* (1882) emphasized the disastrous results, for individuals, of going into debt.

Part Two: New Territories

The principal source for the Gold Rush chapters is Gurdon P. Lester, "'Enrout to Pikes Peak': The 1860 Travel Diary of Gurdon P. Lester," ed. Virginia Foote Anderson and Richard

E. Jensen, which appeared in the Spring 2000 issue of *Nebraska History.* Other nineteenth-century sources include *Oregon and California in 1848,* vol. 2, by J. Quinn Thornton and the *Clayton County Journal* for 1859 and 1860. Among more recent sources, Robert Leaman Brown's *The Great Pikes Peak Gold Rush* focuses on the gold rush narrative itself, while Merrill J. Mattes's *The Great Platte River Road: The Covered Wagon Mainline Via Fort Kearny to Fort Laramie* places the goldseekers' route in a broader historical context, and Elliott West's *The Contested Plains: Indians, Goldseekers, and the Rush to Colorado* underscores the destructive effect of the goldseekers' presence upon the lives and livelihoods of the Native Americans.

Metaphorically speaking, Amorette entered new territory when she became a mother. The chapter "Amorette's Baby" seeks to chart the territory by drawing on sources such as Catharine Beecher's *Treatise on Domestic Economy, for the Use of Ladies at Home, and at School* (1848); Sylvia D. Hoffert's *Private Matters: American Attitudes Toward Childbearing and Infant Nurture in the Urban North, 1800-1860*; *"A Secret to Be Burried": The Diary and Life of Emily Hawley Gillespie, 1858-1888,* edited by Judy Nolte Lensink; and *Brought to Bed: Child-Bearing in America, 1750-1950,* by Judith Walzer Leavitt. By way of comparison, *The Common Sense Book of Baby and Child Care,* by Benjamin Spock, M. D. (1946), treats the same subject from a twentieth-century viewpoint.

Part Three: Seth's Story

Two sources documenting Seth's war experiences have been used again and again in putting his story together. One is his file of Military Records (Record Group 94, held at the National Archives in Washington, D. C.). Its contents include copies of the muster rolls, which report (not always accurately) his whereabouts month by month. The other is the chapter on the Eighth Kansas Volunteer Regiment in *Military History of Kansas Regiments During the War for the Suppression of the Great Rebellion* by William S. Burke (1868, 1870). Burke's frequent use of the pronoun "we" in this chapter indicates that he was a member of the Eighth Kansas.

Almost all the letters and reports of Civil War officers that appear in Part Three are quoted from *The War of the Rebellion: A Compilation of the Official Records of the Union and Confederate Armies,* edited by Lt. Col. Robert M. Scott and Lt. Col. Henry M. Lazelle (1888). (The title is often abbreviated as *Official Records,* or *O. R.*.) The one exception is the letter of Sol Washer, Adjutant of the Eighth Kansas, to Seth Foot, which I possess.

As a geographical source for Seth's service in the western territories, *Fort Laramie and the Pageant of the West* by Leroy Hafen and Francis Marion Young (1938) is memorable for its poetic descriptions of mountains and rivers. I have borrowed two images from these authors: the image of "the shining Laramie Peak" and the one (which Hafen and Young in turn had borrowed) of the mingling of the Plattte and Laramie rivers. James Larson's memoir, *Sergeant Larson, 4th Cavalry* (1935) not only tells the tale of the stolen horse but also tends to confirm Seth's presence at Elk Mountain. *The Civil War in the Western Territories,* by Alvin M. Josephy, Jr., provides further confirmation, and J. J. Hollingsworth's letter about the great numbers of wild animals at Elk Mountain (now preserved in the Wyoming State Archives), captures a vanished scene.

The story of Seth's part in the battle of Chickamauga draws upon Peter Cozzens' *This Terrible Sound: The Battle of Chickamauga.* The portrait of Colonel Heg is based upon *The Civil War Letters of Colonel Hans Christian Heg,* edited by Theodore

C. Blegen. Blegen's introduction includes a character sketch of Colonel Heg and a moving report of his wounding and death. Lieutenant Colonel James L. Abernathy's description of the back-and-forth action on the first day of Chickamauga appears in the *Official Records.*

The record of Seth's participation in the battle of Missionary Ridge depends not only on the muster rolls but more specifically on the crucial report of Brigadier General August Willich (December 31, 1863) in the *Official Records.* Equally important in documenting the fact of Seth's wounding is a single card, called his "carded medical record," which was copied from Colonel John A. Martin's List of Casualties, 8th Kansas (November-December 1863). This card is housed in Seth's Military File at the National Archives.

A brief biography of General Willich appears in Whitelaw Reid's *History of the State During the War, and the Lives of Her Generals,* which is Volume 1 of his *Ohio in the War: Her Statesmen, Her Generals, and Soldiers* (1863). L. Van Loan's *Grape and Canister: The Story of the Field Artillery of the Army of the Potomac, 1861-65* contains an instructive survey of Civil War ammunition.

Much of the material on the probable care and treatment of Seth's wound was gathered from the sections on the battles for Chattanooga in *The Medical and Surgical History of the War of the Rebellion,* Appendix to Part I, edited by Surgeon General Joseph K. Barnes, Assistant Surgeon J. J. Woodward, and Assistant Surgeon George Otis. This nineteenth-century book was reprinted in 1990 as *The Medical and Surgical History of the Civil War.* George Worthington Adams' *Doctors in Blue: The Medical History of the Union Army in the Civil War* provides specific information about the often fatal complications arising from infected wounds. An anecdote in Shelby Foote's Volume

2 of *The Civil War, A Narrative,* serves as a comparison with Seth's case, as does an anecdote recounted by John W. Liddell in Harold Elk Straubing's book, *In Hospital and Camp: The Civil War Through the Eyes of Its Doctors and Nurses.*

The soldiers' letters quoted in James M. McPherson's *For Cause and Comrades* led McPherson to believe that support for the Union and loyalty to their comrades in arms were the Northern men's principal motives for continuing to serve, despite immense hardships. At the same time there were men in Iowa who did not look favorably upon service in the army. The closing of Iowa's western border, in order to prevent such men from escaping the draft, is described in Benjamin Gue's *History of Iowa,* Volume 3 (1903).

PART FOUR: WIDOW AND SON

Almost all the facts concerning the settlement of Seth's estate come from the Will and Settlement file of Seth Foot, which has been described in Part One. The explanation of the "set-aside" fund for widows and children in Iowa is drawn from Ruth A. Gallaher, *Legal and Political Status of Women in Iowa* (1918).

The conditions an applicant for a widow's pension had to meet are spelled out in *Civil War Firsts: The Legacies of America's Bloodiest Conflict* by Gerald S. Henig and Eric Niderost (2001). From that point on, my narrative draws from Seth Foote's Military File, Pension Application File and Volunteer Service File, which are housed at the National Archives.

The Catalogue of the Upper Iowa University lists particulars of Dell's college record, as well as suggesting the weekly rate Amorette may have charged her boarders. A manuscript, "Susan Angelina Collins," which is preserved in the Upper Iowa University archives, seems to be the best record available of Collins's early life. It is accompanied by a photograph of Collins. John D. Clinton's article, "Susan, A Fayette, Iowa Star," in the *Upper Iowa Press* (1975) speaks of her later years. Anecdotes about Fayette's Methodist church and its organ come from an article by Edna Luce in *Centennial 1874-1974,* housed in the archives of the Fayette Public Library.

Background material on homesteading in Dakota Territory is provided by Suzanne Julin, who compiled *Collections of the State Historical Society of South Dakota.* Samuel W. Glenn's *Report on the Tornado of August 28, 1884, near Huron, Dakota* describes in detail the visitation of the white tornado. The forty-six Declarations of Intent which Dell penned as Deputy County Clerk are housed in the South Dakota State Archives. The notice advertising Dr. J. W. Primm's services as a "homoeopathist" appears in the *Dakota Huronite* (26 July 1883).

Part Five: All the Word "Mother" Implies

The history of homeopathy in nineteenth-century America and its guiding principles are found in *Homeopathy in America: the rise and fall of a medical heresy,* by Martin Kaufman (1971). William Harvey King's *History of Homoeopathy and Its Institutions in America* (1905) contains a short biography of "Dellizon Arthur Foote" and identifies Dell as an alumnus of the Chicago Homoeopathic Medical College. *The Rise of Industrial America* by Page Smith describes living conditions for immigrants to Chicago in the late nineteenth century.

The historical context for Amorette's work at the Chicago Training School is provided by two writers—one of them the founder of the school and the other a trained historian, writing a century later. Lucy Rider Meyer's *Deaconesses, Biblical, Early Church, European, American* (1892) tells how the Training School came into being and how the Methodist deaconess movement emerged. Mary Agnes T. Dougherty's *My Calling to Fulfill: Deaconesses in the United Methodist Tradition* (1997) carries the history of the movement up to 1996. It was Dougherty who reached the conclusion, based on her research, that the majority of women who joined the deaconess movement did so because they felt "called to the work by God."

A detailed description, by the Reverend J. C. Jackson, of the Chicago Training School's building appeared in *The Message: Supplement,* April 1887. The Chicago Training School Announcement, published yearly, spelled out rules, policies and duties as they affected students. Isabella Thoburn's thoughtful comments on a deaconess's life and work are quoted in Rider Meyer's book. *Methodism in Omaha: A History of the Methodist Episcopal Church in Omaha and Suburbs* (1895) by the Reverend James Haynes describes the meeting at which Dell spoke of the Chicago deaconess program.

For the chapters on Amorette's service at Talladega College the works of several historians were consulted. *Talladega College: The First Century,* by Maxine D. Jones and Joe M. Richardson, is a full-length history, indispensable to readers who wish to follow the life of this college from its birth immediately after the Civil War to its status at the time of the book's publication (1990). Clara Merritt De Boer's book, *His Truth Is Marching On: African Americans Who Taught the Freedmen for the American Missionary Association 1861-1877* singles out Talladega as the first college in the United States to be open to "colored" students. Ida B. Wells's report on lynchings in Alabama in 1893 appears in *A Red Record: Tabulated Statistics and Alleged Causes of Lynchings in the United States, (1892-1893-1894-1895).*

The most specific source for the first two Talladega chapters is Minutes of the Prudential Committee, 1892-95, which is held in the Talladega College Archives. Lists of faculty members at Talladega from 1892 to 1898 come from the *American Missionary* for those years. The local newspaper that reported the blizzard of 1895 and the death of President De Forest was *Our Mountain Home.*

The chapter on a student protest at Grinnell College in 1942 draws historical facts from *Grinnell College in the Nineteenth Century: From Salvation to Service* (1997) by Joseph Frazier Wall. In *From Slavery to Freedom: A History of American Negroes* (1956) John Hope Franklin explains the policy, adopted by the U. S. Army in 1940, that allowed "Negroes" (as African-Americans were then called) to be admitted to Officer Candidate School. John Schollte Nollen's history, *Grinnell College* (1953) reports the arrival of the Officer Candidate School at Grinnell but does not mention the presence of the "Negro" candidates.

The health center at Clifton Springs, New York, where Amorette went after leaving Talladega, is still in existence. An article seen on the internet, "Clifton Springs Hospital and Clinic" included a photograph and drawing of the hospital and its grounds as they appeared in the nineteenth century.

Amorette's voice is heard in the Annual Reports of the General Board of Managers of the Woman's Home Missionary Society for the years 1901 to 1904 and again in 1906. The photograph, "Southern California Conference Deaconess Home," first appeared in *An Exhibit of the Industrial Homes, Missions, and Deaconess Homes of the Woman's Home Missionary Society of the Methodist Episcopal Church.*

An article, a photograph and a map—all of them on the internet—help to place the Deaconess Home geographically. The article "Los Angeles in the 1900s: La Grande Station" is by George Garrigues. The photograph, "A Visit to Old Los

Angeles," comes from the collection of Brent C. Dickerson. The map, an enlargement of one small portion of Worthington Gates's map of Los Angeles in the early twentieth century, was also found on Garrigues' website.

The visit of Susan Collins to the Southern California Conference Deaconess Home was described for readers of the *California Christian Advocate* (31 October 1901) by Abbie Mills. The deaconess who employed the phrase ". . . all the word 'mother' implies" in the *California Christian Advocate* (16 April 1903) is not identified by name. The identity of M. Eleanor Rich as Amorette's niece was established by her gravestone in the Grandview Cemetery in Fayette and by a remark in Dell's 1923 diary. Will A. Knighten's complaints about the Hewitt Street house appeared in the *Southern California Conference Journal* for 1906 and 1907.

The death certificate issued by the state of Nebraska reports that the cause of Amorette's death was a cerebral hemorrhage. The skull fracture she suffered in her fall is listed as a "contributory" cause. In the *Deaconess Advocate's* announcement, the Bible verse paraphrased is Revelation 14:13: "And I heard a voice from heaven saying unto me, Blessed are the dead which die in the Lord from henceforth: Yea, saith the Spirit, that they may rest from their labors; and their works do follow them."

Printed in the United States
146218LV00001B/2/P

9 781587 901645